Frontiers in Ceramic Science

(*Volume 3*)

Applications of Advanced Ceramics in Science, Technology, and Medicine

Edited by

K. Chandra Babu Naidu

Department of Physics
GITAM Deemed to Be University
Karnataka
India

&

N. Suresh Kumar

Department of Physics
JNTUA College of Engineering
Andhra Pradesh
India

Frontiers in Ceramic Science

Volume # 3

Applications of Advanced Ceramics in Science, Technology, and Medicine

Editors: K. Chandra Babu Naidu and N. Suresh Kumar

ISSN (Online): 2542-5269

ISSN (Print): 2542-5250

ISBN (Online): 978-981-14-7819-2

ISBN (Print): 978-981-14-7817-8

ISBN (Paperback): 978-981-14-7818-5

need for a court order if at any point you breach any terms of this License Agreement. In no event will any delay or failure by Bentham Science Publishers in enforcing your compliance with this License Agreement constitute a waiver of any of its rights.

3. You acknowledge that you have read this License Agreement, and agree to be bound by its terms and conditions. To the extent that any other terms and conditions presented on any website of Bentham Science Publishers conflict with, or are inconsistent with, the terms and conditions set out in this License Agreement, you acknowledge that the terms and conditions set out in this License Agreement shall prevail.

Bentham Science Publishers Pte. Ltd.
80 Robinson Road #02-00
Singapore 068898
Singapore
Email: subscriptions@benthamscience.net

CONTENTS

PREFACE

Advanced ceramics are specified as the new class of ceramic materials made up of high purity synthetic chemicals. In recent years these ceramics gained much research attention in various applications due to their excellent performance. These advanced ceramics are composed of oxides, carbides and nitrides. Ceramic composites which include both the combinations of oxides and non-oxides Ceramic materials are the different class of materials composed of non-metallic stable and inorganic materials. They are brittle, electrically insulated and thermally insulated materials prepared with combination of more than single element. The ceramic materials will have distinct properties due to their grain boundaries in the materials undergo misalignments with nearby grains and their variations in the structure and quality in the perfections and contributions along with shape and size and material internal stress which they are exposed. For the few decades' people have observed the tremendous increase in the properties like ferroelectric, ferromagnetic, pyroelectric, dielectric and magnetoresistance, superconducting and gas sensing applications. These ceramic materials also became main materials for the advanced technologies like energy transformation storage and supply and also in the manufacturing, medical field technology and transportation, information technology *etc*. Hence, awareness and knowledge about ceramics and their derived nanomaterials with conceptual understanding are important for the advanced material community.

Advanced Ceramics and Applications in Science, Technology and Medicine explores the various advanced ceramic materials and their down to earth applications in distinct fields such as actuators, energy storage, environmental, 3D printing, electronics, biomedical and EMI shielding. This book provides an overview of the structural and fundamental properties, synthesis strategies and versatile applications of advanced ceramic materials and their composites. This book will be beneficial for students, research scholars, faculty members, R & D specialists working in the area of material science, solid-state science, chemical engineering, power sources and renewable energy storage fields and nanotechnologists. Based on thematic topics, the book contains the following 12 chapters:

Chapter 1 discuss briefly about the importance of piezoelectric actuators and also the principle of piezoelectric actuators. Besides, it is outlined on various advanced ceramic materials for the applications in piezoelectric actuators. In addition to this, some actuator operating methods like finite element method, topology optimization method *etc*. are also discussed.

Chapter 2 presents briefly an introduction of ceramics, different types of ceramic materials and role of ceramic materials in the evaluation of supercapacitors. Further, it is focused on different advanced ceramic electrode materials for supercapacitor applications. In addition to this some fabrication techniques like hydrothermal technique, molten salt technique, solution precursor flame spray *etc*. is discussed as well.

Chapter 3 focused on future perspective of magnetocaloric effect in refrigeration process and the isothermal entropy and adiabatic variation are the main properties to confine the refrigeration process. Also, the magnetic materials such as Mn doped Fe alloys and rare earth elements such as Gd, La, Ce *etc*. mixed composites and their properties are discussed.

Chapter 4 discusses the efficiency of the thermoelectric power generation depends on the figure of merit of the materials and also on the low thermal conductivity and high electrical conductivity. Several ceramic materials and their applications in the field of power generation are also discussed.

Chapter 5 reviews different ceramic materials including their electromagnetic parameters. The electromagnetic absorption property of the ceramic composites and pure ferrites are discussed. Furthermore, the applications of the advanced ceramics in defense systems, microwave communications and biomedical fields are summarized.

Chapter 6 briefly presents the discussion on the effect of electromagnetic radiation on the electronic goods, human health and defense system was elaborated. Furthermore, various ceramic materials were introduced for reducing the electromagnetic radiation pollution thereby microwave absorption process. It is focused on the parameters like magnetic loss and dielectric loss for each ceramic material. Subsequently, the applications of microwave absorbers in various fields were elucidated.

Chapter 7 concentrates on electromagnetic wave interference mechanism and the effected parameters to find the strength of shielding. Also discusses the derived material such as ferrites and its composites, carbon-based materials for shielding the EM interference. In addition, ferrite and polymer composites especially conducting polymer composites have been discussed.

Chapter 8 discusses the presence of intrinsic polarizations into the ferroelectric materials helps them to get polarized easily on the application of the electric field. Also, the mechanism and applications of ferroelectric materials ($Bi_{3.25}La_{0.75}Ti_3O_{12}$, Barium hafnium titanate, $La_3Ni_2NbO_9$ *etc.*) in different fields like ferroelectric memory devices, electro caloric devices, magneto electric devices, DRAM capacitors *etc.* have been discussed.

Chapter 9 reviews the transport properties of semiconducting glasses along with amorphous properties of various materials. In addition, some of the synthesis techniques such as thermal evaporation, chemical vapour deposition, melt quenching *etc.* have also been discussed.

Chapter 10 presents the applications of 3D printing ceramic materials in various fields. Also discusses the advantages of 3D printing over conventional techniques.

Chapter 11 focused on bio ceramics and preparation of bio ceramics by advanced 3D printing technology. Besides, different inorganic materials used to print bio ceramics such as alumina, zirconia, Leucite, lithium disilicate and mica-based ceramics for different applications in dentistry and orthopedics are also discussed.

Chapter 12 presents briefly the origin of ceramics, its advantages and the historical back ground of the ceramic materials and the classification of ceramics. Moreover, antimicrobial activity and the antimicrobial applications of advanced ceramics are summarized.

KEY FEATURES

- Overviews on advanced ceramic materials and their derived composites.
- Coverage on basic research and application approaches.
- Addresses a wide range of applications in actuators/sensors, energy conversion and storage, 3D printing, antimicrobial, EMI shielding and microwave absorbers.
- Explores challenges and future directions of advanced ceramic materials.

K. Chandra Babu Naidu
Department of Physics
GITAM Deemed to Be University
Karnataka
India

&

N. Suresh Kumar
Department of Physics
JNTUA College of Engineering
Andhra Pradesh
India

LIST OF CONTRIBUTORS

A. Manohar	Department of Materials science and engineering, Korea University, Republic of Korea
A. Mallikarjuna	Department of Physics, GITAM Deemed to be University, India
Abdullah M. Asiri	Department of Chemistry, King Abdulaziz University, Saudi Arabia
Anish Khan	Department of Chemistry, King Abdulaziz University, Saudi Arabia
Adolfo. Franco	Instituto de Física, Universidade Federal de Goiás, Goiânia, Brazil
B. Venkata Shiva Reddy	Department of Physics, GITAM Deemed to be University, India
B. Kishore	Department of Mechanical Engineering, GITAM Deemed to be University, India
B.V. Rama	Department of Physics, GITAM Deemed to be University, India
D Ravinder	Department of Physics, Osmania University, India
K. Chandra Babu Naidu	Department of Physics, GITAM Deemed to be University, India
K. Venkata. Ratnam	Department of Chemistry, GITAM Deemed to be University, India
K.V. Niranjan	Department of Physics, GITAM Deemed to be University, India
K.V. Ramesh	Department of Electronics and Physics, GITAM Deemed to be University, India
Khalid Mujasam Batoo	King Abdullah Institute For Nanotechnology, King Saud University, Saudi Arabia
M. Prakash	Department of Physics, Sri Krishnadevaraya University, India
M.S.S.R.K.N. Sarma	Department of Physics, Andhra University, India
N. Suresh Kumar	Department of Physics, JNTUA, India
N.V. Krishna Prasad	Department of Physics, GITAM Deemed to be University, India
Prasun Banerjee	Department of Chemistry, GITAM Deemed to be University, India
R. Padma Suvarna	Department of Physics, JNTUA, India
R.J.S. Lima	Instituto Federal de Educacao Ciencias e Tecnologia de Goias, Brazil
S. Ramesh	Department of Physics, GITAM Deemed to be University, India
Sannapaneni Janardan	Department of Chemistry, GITAM Deemed to be University, India
Thiago E.P. Alves	Unidade Academica de Fisica, Universidade Federal de Campina Grande, Brazil
U. Naresh	Department of Physics, BIT Institute of Technology, India

CHAPTER 1

Advanced Ceramics for Piezoelectric Actuators

N. Suresh Kumar[1], R. Padma Suvarna[1,*], K. Chandra Babu Naidu[2,*] and Khalid Mujasam Batoo[3]

[1] *Department of Physics, JNTUA, Anantapuramu-515002, A.P, India*

[2] *Department of Physics, GITAM Deemed to be University, Bangalore-562163, Karnataka, India*

[3] *King Abdullah Institute For Nanotechnology, King Saud University, P.O. Box 2455, Riyadh, 11451, Saudi Arabia*

Abstract: Nowadays, researchers concentrated on the advancements of new types of precision actuators to satisfy the increasing demand for high precision positioning technology. Owing to fast response, high precision, and compact structure, the piezoelectric actuators attracted much attention. In the advancement of science and technology, the ceramics play a vital role. In this chapter, we mainly focussed on applications of piezoelectric actuators in different fields such as nano metrology, and industries. Even, the different advanced ceramics for piezoelectric actuators were discussed.

Keywords: Acoustic actuators, Dielectric properties, Piezoelectric effect.

1.1. INTRODUCTION

In many areas such as MEMS, NEMS, nano metrology and biological engineering, the demand for ultra precision and nano plating technologies have been increased along with the development of science and engineering [1 - 3]. Accordingly, the significance of nano positioning systems has been increased in many industrial and research areas [4, 5]. The conventional actuators like DC-motors, AC-motors, hydraulic motors, *etc.*, can fulfill the above requirements in the large scale positioning systems. However, the resolution of those actuators is not high [6 - 8]. Besides, the electromagnetic interference affects the performance of the conventional electromagnetic motors; no electromagnetic resonance can be generated by the piezoelectric actuators and even cannot be affected by electromagnetic interference. Therefore, in recent years, the researchers have

---

* **Corresponding author: Dr. K. Chandra Babu Naidu:** GITAM Deemed To Be University-Bangalore Campus, Bangalore-562163, Karnataka, India; E-mails: chandrababu954@gmail.com & padmajntua@gmail.com

focussed on introducing new kinds of actuators to meet the demands of nano-positioning systems. In this connection, piezoelectric actuators gained much attention owing to their quick response, simple structure, self-locking, *etc.* In addition, these actuators exhibit the resolution in nanoscale and suitable for nano positioning systems. Till now these piezo actuators are used in numerous fields such as optical lens systems [9, 10], aerospace [11, 12], and atomic force microscopy [13, 14], *etc.* In general, the piezoelectric materials which can be utilized as actuators are also called piezoelectric ceramics which consist of a specific crystal structure [15, 16]. They exhibit some unusual properties which make them special when compared to others. Those properties are due to a direct piezoelectric effect and indirect piezoelectric effect. In the case of the direct piezoelectric effect, these materials generate an electric charge by subjecting with mechanical stress. Similarly, in the inverse piezoelectric effect, the stain is developed in the same material by the application of the electric field. The direct effect is used in piezoelectric sensors while the inverse effect is used in actuators [17, 18]. In this chapter, we focussed on the inverse effect of piezoelectric ceramics and their applications in actuators.

1.2. ADVANCED MATERIALS FOR PIEZOELECTRIC ACTUATORS

In continuation with the development of designing the new type of piezoelectric actuators, Markus Flossel *et al.* [19], introduced a new module design based on the LTCC (Low-Temperature Co-fired Ceramic)/PZT (Lead Zirconate Titanate) multilayer ceramics for the actuator applications. This new technology involves the lamination of PZT ceramic plate sintering with the green layer LTCC followed by sintering to attain a sensor or actuator segment. (Fig. **1.1**) of the reference [19], represents the design of LTCC/PZT. Further, they reported that the functioning of the designed module is estimated by measuring dielectric properties and recording the hysteresis loops. In addition, they measured the deflection and reported that at a resonance frequency of 19.7 Hz, the prepared ceramic module exhibits the maximum deflection of 130×10^{-6} m which is in static mode, whereas in dynamic mode the same module shows the deflection of 550×10^{-6} m. Furthermore, through the metal die casting technique, the prepared modules are combined with Al (aluminium) matrix for further enhancement of the properties of the designed module. However, this new technology combines the microsystem technology of LTCC and piezoelectric technology of PZT which allows great improvement in practical applications such as electronic control, sensing and actuation.

Table 1.1. Advanced ceramic materials for piezoelectric actuator applications.

S.No.	Material	Synthesis Method	d_{33} (CN^{-1})	Electro Strain (in %)	Applications	Ref.
1	LTCC (Low-Temperature Cofired Ceramic)/PZT (Lead Zirconate Titanate) multilayer ceramics				Actuator, sensor, electronic control	[19]
2	$0.91K_{1/2}Bi_{1/2}TiO_3$–0.09 ($0.82BiFeO_3$-$0.15NdFeO_3$-$0.03Nd_{2/3}TiO_3$ ceramics	conventional solid-state and mixed route		0.16	actuator	[20]
3	$(Na_{0.48-x}K_{0.48-x}Li_{0.04})Nb_{0.89-x}Ta_{0.05}Sb_{0.06}O_3$-$xSrTiO_3$ ceramics	tape casting method and solid-state reaction method	650		audible sound device applications	[21]
4	$Pb(Mn_{1/3}Nb_{2/3})\,0.07(Ni_{1/3}Nb_{2/3})\,0.10(Zr_{0.5}Ti_{0.5})0.83O_3$ ceramics	conventional mixed oxide synthesis technique	386		piezoelectric speakers	[23]
5	$Ba_{1-x}Ln_{2x/3}Zr_{0.3}Ti_{0.7}O_3$ ceramics	solid state reaction technique		0.087	energy storage and actuator applications	[25]
6	periodically orthogonal poled method to barium titanate based ceramics			0.36	large actuation at low temperature and low frequency	[27]
7	$(Zr_{0.49}Ti_{0.51})_{0.94}Mn_{0.014}Sb_{0.02}W_{0.014}Ni_{0.02}O_3$ ceramics	calcination technique	278		micromechatronic devices	[28]
8	Li-doped BNT ($Bi_{0.5}Na_{0.5}TiO_3$) ceramic materials	solid state reaction technique	600		actuator applications	[30]

Amir Khesro *et al.* [20], synthesized lead-free $(0.91K_{1/2}\,Bi_{1/2}TiO_3$-$0.09$ $(0.82BiFeO_3$-$0.15NdFeO_3$-$0.03Nd_{2/3}TiO_3)$ ceramics for actuator applications with good thermal stability and fatigue resistant. These samples were prepared through conventional solid-state and mixed routes. The outcomes revealed that the prepared ceramics exhibit high electro strain of 0.16% at 6000 V/mm with a 10% variation from room temperature to 448 K. Therefore, the ceramics show good stability in this temperature range. Also, they designed a multilayer actuator through cobalt firing with Pt internal electrodes. These multilayer actuators exhibit thermal stability up to 300°C with a small variation of 15%. Further, the same ceramics show high fatigue resistance with both multi and mono layers. Hence, these advanced ceramics show the superior properties and are potential candidates for actuators when compared to all other lead-free ceramics. The following Table **1.1** represents some of the advanced ceramic materials used for piezoelectric actuator applications.

Using the multilayer technology, Su *et al.* [21], prepared the piezoelectric acoustic actuators. For this purpose, they prepared non-stoichiometric lead-free $(Na_{0.48-x}K_{0.48-x}Li_{0.04})\ Nb_{0.89-x}Ta_{0.05}Sb_{0.06}O_3\text{-}xSrTiO_3$ (NKLNTS-xST) (where x=0 and x=0.007) ceramics for the piezoelectric actuator applications by using the tape casting method and solid-state reaction method. They investigated the phase and domain structures of the prepared ceramics. The X-ray diffraction studies revealed the coexistence of the tetragonal and orthorhombic phases in x=0.007 compound. Further, the TEM and SEM images provide the information about the morphology of the prepared materials. We know that in the enhancement of piezo and ferroelectric properties, the domain structures have an important role. Again, for x=0.007 compound in NKLNTS-xST ceramics, the stripe and herringbone domain morphology were observed which increase the wall density. In general, when the tetra and orthorhombic phases are coexisted, the domain is originated in nanoscale dimension which can improve the electrical properties. Furthermore, the NKLNTS-xST ceramics for x=0.007 compound exhibit the high piezoelectric coefficient of 650 pCN^{-1}. Therefore, the comparative studies evidenced that the piezoelectric acoustic actuator with nonstoichiometric lead free NKLNTS-xST (where x = 0.007) ceramic material exhibits high coexistence of orthorhombic-tetragonal phases, stripe & herringbone morphology, nanoscale dimensions, large piezoelectric co-efficient and high electric properties. Therefore, the NKLNTS-xST (where x = 0.007) piezo acoustic actuator is a potential candidate for lead-free audible sound device applications.

Bruno *et al.* [22], compared the performances of one PMN-PT and 8-PZT ceramics for actuator applications. They observed that all the materials show high values of charge coefficients and also the enhancement in charge coefficients with respect to the field strength up to saturation point. After this decrement in charge, coefficients take place. They concluded that PMN-PT ceramic material shows high d_{33} co-efficient than other materials. Besides, the same materials exhibit the low Curie temperature and low coercive field strength. Despite of this, the PZT ceramic materials exhibit high operating field voltage up to 50% of E_c with an extension up to 95%. Furthermore, these materials show more than 90% of retention after million operation cycles. Juhyun Yoo [23], investigated the dielectric and piezoelectric properties of the PNN-PMN-PZT Ceramics sintered at low temperatures. In recent years, the piezoelectric actuators are also used in piezoelectric speakers. But still, the scientists are searching for the new materials for piezoelectric actuators which can be used in piezoelectric speakers. Yoo [23], proposed the PNN-PMN-PZT ceramics and reported that with the help of conventional mixed oxide synthesis technique $Pb(Mn_{1/3}Nb_{2/3})_{0.07}\ (Ni_{1/3}Nb_{2/3})_{0.10}$ $(Zr_{0.5}Ti_{0.5})_{0.83}O_3$ composition ceramic materials were prepared. In addition, Yoo

[23], studied how the microstructure and piezoelectric properties affected with ZnO and Nb_2O_5. The enhancement of Nb_2O_5 content increases the tetragonality of the prepared ceramics. Moreover, for the same content which was sintered at the temperature of 940°C exhibits the high relative permittivity of 1667, K_p value of 0.661, Q_m is 1285 and d_{33} coefficient of 386 PC/N. Hence, these outcomes evidenced that the prepared ceramics are capable of serving low loss actuators which are useful for piezoelectric speakers. Jie Deng *et al.* [24], proposed a highly sensitive planar piezoelectric actuator with hybrid bending-bending operating mode. The proposed actuator consists of four transducers which can bend vertical and horizontal directions dependently. Further, it can be operated in two different modes which are dynamic and quasi-static mode. For large scale measurement with high speed, the dynamic mode is much suitable in which the hybrid bending is used. Whereas, the quasi static mode is suitable for small scale measurements with very high resolution and horizontal bending played a key role in this mode of operation. The (Fig. **1.1**) of the reference [24] represents the structure of the proposed planar actuator.

In addition, they reported that FEM (finite element method) is utilized to explain the operating principle of the actuator along with simulation. Further when the device is operated at 400 voltage and 11.86 kHz frequency, in dynamic mode it displayed a maximum linear speed of about 19.8 mm/s and rotatory speed of 0.266 rad/s. But, in the quasi static mode, it exhibited the same values in nano range of the order of 16 nm and 198 nano radians. Hence, the planar actuator can find the applications in various areas like scanning, transporting *etc.*, with very high resolution. By using solid state reaction technique, Ghosh *et al.* [25], prepared $Ba_{1-x}Ln_{2x/3}Zr_{0.3}Ti_{0.7}O_3$ ceramics where Ln=La, Nd, Sm, Eu, and Sc and x = 0.2-1.0 for energy storage and actuator applications. They reported the effect of rare elements on the efficiency of the energy storage and electro-strictive coefficients of the prepared ceramics. Besides, the outcomes of the investigations evidenced that the compound (2-4 volume %) exhibit the improved energy density around 2.6 times the pure BZT ceramic material. The strain (S) of 0.087% and large electro-strictive coefficients (M_{11} & Q_{11} of 0.16 x 10^{-19} m^2/V^2 & 6 x 10^{-2} m^4/V^2) were noticed. Therefore, the substitution of the rare earth elements causes the improvement in the efficiency of the prepared ceramics due to relaxor phase and slim hysteresis loop. Hence, these properties indicated that the lead free BLnZT ceramics are potential candidates for actuator and energy storage applications.

In the advancement of piezoelectric actuator technology, the Pommier-Budinger *et al.* [26], developed piezoelectric actuator based de-icing system, in which they introduced a computational method to estimate the current and voltage required to induce the first ice delamination/cracks. They reported that the proposed method is also helpful to estimate the frequency range with respect to the performance of the de-icing system. Hence, the piezoelectric actuators also play vital role in de-icing systems which regulate the power consumption in such systems. Generally, the high lead-concentration leads to toxicity problem and the piezoelectric actuators based on PZT-ceramics suffer from small output strains of the order of 0.1 to 0.15% to overcome the problems of these PZT based actuators. Qiang zhong Wang and Faxin Li [27], applied periodically orthogonal poled (POP) method to barium titanate (BT) based ceramics and also designed a multilayer actuator which can exhibit large actuation strain in plane poled region of the order of 0.36% through reversible domain switching. The mechanism of the reversible domain switching can be understood by the structure (tetragonal) of the proposed ceramics. It is arising due to the large internal compressive-stresses in both plane-poled and thickness-poled regions. From the outcomes, they reported that along with the period direction, the non-uniform actuation strain (0.22-0.36%) is observed in the proposed actuator. Nevertheless, in-plane poled regions, the layers of the proposed actuator are bonded together which lead to uniform actuation strain of about 0.34% at 0.1 Hz under 2 kV/mm. In addition, the actuation strain is stable and uniform even after 20000 cycles of operation. But, it decreases very quickly with increasing frequency. Therefore, the proposed POP-BT ceramics are very useful when lead-free ceramics are compulsory at low temperature actuation. However, these are not useful at high temperatures because of low Curie temperature of the BT ceramics. So, the POP-BT ceramic actuators are very useful for attaining large actuation at low temperature and low frequency. Dariusz Bochenek *et al.* [28], investigated the electro-physical properties of Mn^{4+}, Sb^{3+}, W^{6+} and Ni^{2+} doped lead zircon ate ceramics of $(Zr_{0.49}Ti_{0.51})_{0.94}$ $Mn_{0.014}$ $Sb_{0.02}$ $W_{0.014}Ni_{0.02}O_3$ synthesized *via* calcination technique followed by free sintering. They reported that the prepared multicomponent ceramic powders exhibit the tetragonal phase with no secondary phases confirmed from X-ray diffraction studies and scanning electron microscope images. Further, for the same samples the observed low values of dielectric loss (at T_r =0.003 and at T_m=0.12) and high values of dielectric permittivity (at T_r = 1350 and at T_m= 20900) at room temperature (T_r) as well as at transition temperatures (T_m= 573K). Besides, the ferroelectric loops exhibit the broad hysteresis loops (as shown in (Fig. **1.1**) of Ref [28].) with large values of spontaneous polarization (P_s = 18.70 $\mu C/cm^2$). In addition, they attained large electromechanical coupling coefficient (K_p=0.48) and d_{33} coefficient of 278 pC/N. With all these results, Bochenek *et al.* [28], concluded that the prepared multicomponent PZT-ceramics with high values of spontaneous polarization, dielectric permittivity and low values of dielectric loss are much suitable for the applications in modern micro-mechatronic devices such as actuators. Xinqi Tian *et al.* [29], designed a new type of U-shaped stepping piezo electric actuator as shown in Fig. (**1.1**) of reference [29] by using four PAUs (piezoelectric actuation units). The designed actuator consists of two pushing units and two clamping units. The coupling motion is suppressed by the clamping units with the help of brackets fixed at the base and sliders connected to the clamping units. In addition to this, the static model is also developed to measure the clamping force and step length of the proposed actuator. The measured values of the actuator were well matched with the outcomes of the static model. Further, the dependence of the output speed on the frequency and voltage is examined. They observed that the linear increment of the output speed with respect to the voltage in the frequency range 0-5 Hz. After this, up to 6.25 Hz the output speed is increased slightly, beyond 6.25 Hz, it is decreased with increasing voltage. The maximum thrust force of the designed model is about 189 N and output speed is 273 μm/s. These investigations suggested that the new actuator can be useful in the areas such as precision feeding machine tool, precision zoom lens adjustment system and pression positioning platform for microscope. Kang *et al.* [30], prepared Li-doped BNT ($Bi_{0.5}Na_{0.5}TiO_3$) ceramic materials *via* solid state reaction technique. They explored the dependence of electro-strain on temperature of the prepared lead-free ceramics. In addition, they designed a piezoelectric actuator with multilayer through gradient doping and compared the outcomes with single layer one. Further, they reported that five-layer composition shows high electro-strain and high thermal stability. Moreover, the high d_{33} coefficient of 600 pm/V at 70°C is noticed at relatively low electric field. The enhancement of the stability and electro-strain can be achieved by composition adjustment in different layers. Hence, this is a new strategy for the actuator applications with good thermal stability. In addition, to measure the vibrational loss in dynamic mode, Moretti *et al.* [31], designed FPEA (fextensional piezoelectric actuator) by applying TOM (Topology Optimization Method). In the proposed actuator, the fixed shape of the materials helps to optimize the topology of the host structure. Therefore, the actuator which is formed through two layers of the piezoelectric materials can serve as an actuator and also a sensor and provides constant gain.

In this, the actuator is coupled with the AVFC (active velocity feedback control law) which modifies the damping matrix with respect to the observations as a function of time. (Fig. **1.1**) of reference [31] indicates the proposed actuator model design. This model is also named as the 'nc' (non-collapsed)-nodes model because piezoelectric ceramic intermediate nodes are not simplified while optimizing the model through analyzing dynamic equation. Further, with the help of Netmark's time integration method, FE (finite element) analysis attained in order to measure the dynamic response of the rectangular nodes. Also, under the influence of transient mechanical load the energy of the response signal of designed actuator (FPEA) is reduced by gradient based optimization technique. Finally, they compared the outcomes of the prepared model with the collapsed node (c-node) model. From the comparison, it is evidenced that the new nc-model is suitable for the impending advancements that anticipates affording an external circuit to the proposed structure pointing to intensity the damping effect of the AVFC.

Nowadays, these advanced ceramics play an important role in modern vehicles *i.e.*, in modern vehicles, these are used as automatic sensors to ensure smart engine management to keep the passengers safe. In addition, these are used in generating great ultrasonic intensities for ultrasonic drilling and cleaning. Besides, these are also used in ultrasonic transmitters and receivers for transmitting the signal, receiving the signal and processing the information. Furthermore, while operating the piezoelectric actuators the deformation takes place in micrometer range. From this, it can also be used as an actuator for pneumatic and hydraulic values, dispensing systems for gases and liquids, micromanipulators *etc*. In addition to this, the piezoelectric actuators are being used in sound emitters, optical fiber adjustment, active vibration damping *etc*. Research is going on to improve the efficiency of the piezoelectric actuators by using some advanced ceramic materials. In this chapter, we have discussed some of those ceramic materials and their actuator applications. The following (Fig. **1.1**) [31] represents piezoelectric rotatory actuator for dynamic mirror deflection of head up display.

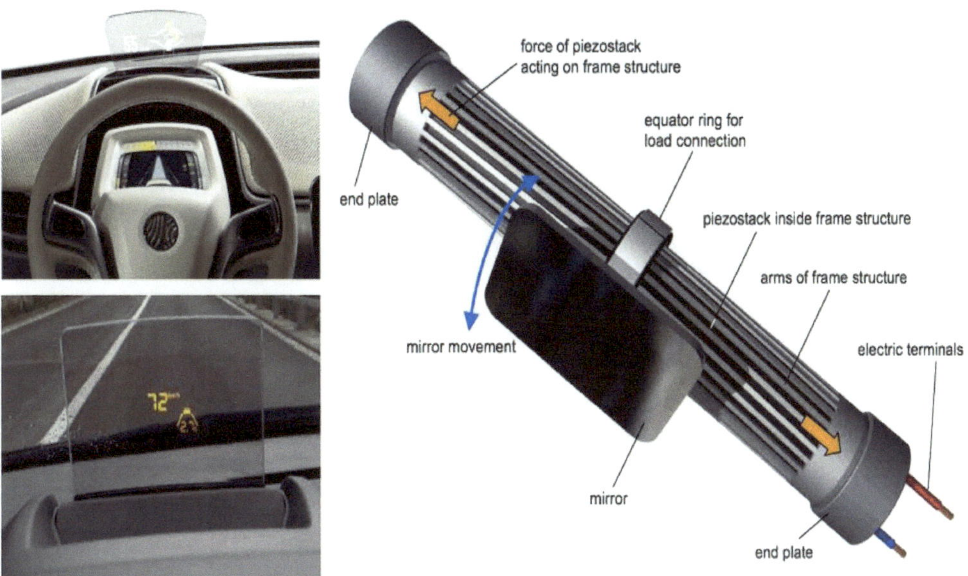

Fig. (1.1). Piezoelectric rotatory actuator for dynamic mirror deflection of head up display.

CONCLUSIONS

We have discussed briefly about the importance of piezoelectric actuators and also the principle of piezoelectric actuators. Besides, we focused on various advanced ceramic materials such as LTCC (Low-Temperature Cofired Ceramic)/PZT (Lead Zirconate Titanate) multilayer ceramics, NKLNTS-xST, PMN-PT, PNN-PMN-PZT, $Ba_{1-x}Ln_{2x/3}Zr_{0.3}Ti_{0.7}O_3$, POP-BT, Li-doped BNT ceramics for the applications in piezoelectric actuators. In addition to this, some actuator operating methods like finite element method, topology optimization method *etc.* are also discussed.

CONSENT FOR PUBLICATION

Not applicable.

CONFLICT OF INTEREST

The authors confirm that this chapter content has no conflict of interest.

ACKNOWLEDGEMENTS

The authors express thankfulness to Dr. P. Sreeramulu, Assistant Professor (English), GITAM, Bangalore for providing English language editing services to this manuscript.

REFERENCES

[1] L. Cheng, W. Liu, Z.G. Hou, J. Yu, and M. Tan, "Neural network based nonlinear model predictive control for piezoelectric actuators", *IEEE Trans. Ind. Electron.,* vol. 62, pp. 7717-7727, 2015.
[http://dx.doi.org/10.1109/TIE.2015.2455026]

[2] F.J. Lin, Y.C. Hung, and S.Y. Chen, "FPGA-based computed force control system using elman neural network for linear ultrasonic motor", *IEEE Trans. Ind. Electron.,* vol. 56, pp. 1238-1253, 2009.
[http://dx.doi.org/10.1109/TIE.2008.2007040]

[3] W.L. Sang, K.G. Ahn, and J. Ni, "Development of a piezoelectric multi-axis stage based on stick-an--clamping actuation technology", *Smart Mater. Struct.,* vol. 16, pp. 2354-2367, 2007.
[http://dx.doi.org/10.1088/0964-1726/16/6/040]

[4] A.J. Fleming, "A review of nanometer resolution position sensors: operation and performance", *Sens. Actuators A Phys.,* vol. 190, pp. 106-126, 2019.
[http://dx.doi.org/10.1016/j.sna.2012.10.016]

[5] X. Zhang, and Q. Xu, "Design, fabrication and testing of a novel symmetrical 3-DOF large-stroke parallel micro/nano-positioning stage, Robot. Cim.-", *Int. Manu.,* vol. 54, pp. 162-172, 2018.
[http://dx.doi.org/10.1016/j.rcim.2017.11.006]

[6] J. Kim, S. Choi, K. Cho, and K. Nam, "Position estimation using linear hall sensors for permanent magnet linear motor systems", *IEEE Trans. Ind. Electron.,* vol. 63, pp. 7644-7652, 2016.
[http://dx.doi.org/10.1109/TIE.2016.2591899]

[7] C. Gradl, A. Plöckinger, and R. Scheidl, "Sensorless position control with a hydraulic stepper drive—concept, compression modeling and experimental investigation", *Mechatronics,* vol. 35, pp.

91-101, 2016.
[http://dx.doi.org/10.1016/j.mechatronics.2016.01.004]

[8]   D. Stoianovici, A. Patriciu, D. Petrisor, D. Mazilu, and L. Kavoussi, "A new type of motor: pneumatic step motor", *IEEE. ASME,* vol. 12, no. 1, pp. 98-106, 2007.

[9]   F. Schneider, J. Draheim, C. Müller, and U. Wallrabe, "Optimization of an adaptive PDMS-membrane lens with an integrated actuator", *Sens. Actuators A Phys.,* vol. 154, no. 2, pp. 316-321, 2009.
[http://dx.doi.org/10.1016/j.sna.2008.07.006]

[10]   Q. Li, L. Liu, and X. Ma, "Development of multi-target acquisition, pointing and tracking system for airborne laser communication", *IEEE Trans. Industr. Inform.,* vol. 15, no. 3, pp. 1720-1729, 2018.
[http://dx.doi.org/10.1109/TII.2018.2868143]

[11]   H. Elahi, M. Eugeni, P. Gaudenzi, F. Qayyum, R.F. Swati, and H.M. Khan, "Response of piezoelectric materials on thermomechanical shocking and electrical shocking for aerospace applications", *Microsyst. Technol.,* pp. 1-8, 2018.
[http://dx.doi.org/10.1007/s00542-018-3856-8]

[12]   H.C. Chung, K.L. Kummari, S.J. Croucher, N.J. Lawson, S. Guo, and Z. Huang, "Coupled piezoelectric fans with two degree of freedom motion for the application of flapping wing micro aerial vehicles", *Sens. Actuators A Phys.,* vol. 147, no. 2, pp. 607-612, 2008.
[http://dx.doi.org/10.1016/j.sna.2008.06.017]

[13]   M.W. Fairbairn, S.R. Moheimani, and A.J. Fleming, "Q control of an atomic force microscope microcantilever: a sensorless approach", *J. Microelectromech. Syst.,* vol. 20, pp. 1372-1381, 2011.
[http://dx.doi.org/10.1109/JMEMS.2011.2168809]

[14]   M.G. Ruppert, S.I. Moore, M. Zawierta, A.J. Fleming, G. Putrino, and Y.K. Yong, "Multimodal atomic force microscopy with optimized higher eigenmode sensitivity using on-chip piezoelectric actuation and sensing", *Nanotechnology,* vol. 30, no. 8, p. 085503, 2019.
[http://dx.doi.org/10.1088/1361-6528/aae40b] [PMID: 30251962]

[15]   J. Ma, Y. Hu, B. Li, Z. Feng, and J. Chu, "Influence of secondary converse piezoelectric effect on deflection of fully covered PZT actuators", *Sens. Actuators A Phys.,* vol. 175, pp. 132-138, 2012.
[http://dx.doi.org/10.1016/j.sna.2011.12.034]

[16]   A. Shafik, and R.B. Mrad, Piezoelectric Motor Technology: A Review. *Nano positioning Technologies.,* C. Ru, X. Liu, Y. Sun, Eds., Springer International Publishing: Switzerland, 2016, pp. 33-59.
[http://dx.doi.org/10.1007/978-3-319-23853-1_2]

[17]   Z. Liu, Z. Yao, X. Li, and Q. Fu, "Design and experiments of a linear piezoelectric motor driven by a single mode", *Rev. Sci. Instrum.,* vol. 87, no. 11, p. 115001, 2016.
[http://dx.doi.org/10.1063/1.4966251] [PMID: 27910463]

[18]   C. Zhu, X. Chu, S. Yuan, Z. Zhong, Y. Zhao, and S. Gao, "Development of an ultrasonic linear motor with ultra-positioning capability and four driving feet", *Ultrasonics,* vol. 72, pp. 66-72, 2016.
[http://dx.doi.org/10.1016/j.ultras.2016.07.010] [PMID: 27479230]

[19]   M. Flössel, S. Gebhardt, A. Schönecker, and A. Michaelis, "Development of a Novel Sensor-Actuato--Module with Ceramic Multilayer Technology", *J. Ceram. Sci. Tech.,* vol. 01, no. 01, pp. 55-58, 2010.

[20]   A. Khesro, D. Wang, F. Hussain, D.C. Sinclair, A. Feteira, and I.M. Reaney, "Temperature stable and fatigue resistant lead-free ceramics for actuators", *Appl. Phys. Lett.,* vol. 109, no. 14, p. 142907, 2016.
[http://dx.doi.org/10.1063/1.4964411]

[21]   H-H. Su, C-S. Hong, C-C. Tsai, and S-Y. Chu, "Domain structure of nonstoichiometric sodium potassium niobate-based ceramics for piezoelectric acoustic actuators", *Ceram. Int.,* vol. 44, no. 4, pp. 3787-3790, 2018.
[http://dx.doi.org/10.1016/j.ceramint.2017.11.163]

[22]   B.B. Poyyathuruthy, A. Fahmy, M. Stürmer, U. Wallrabe, and M.C. Wapler, "Properties of

piezoceramic materials in high electric field actuator applications", *Smart Mater. Struct.,* vol. 28, p. 015029, 2018.

[23]   J. Yoo, "High Dielectric and Piezoelectric Properties of Low-Temperature Sintering PNN-PMN-PZT Ceramics for Low-Loss Piezoelectric Actuator Application", *Transactions on Electrical and Electronic Materials,* vol. 19, no. 4, pp. 249-253, 2018.
[http://dx.doi.org/10.1007/s42341-018-0045-5]

[24]   J. Deng, Y. Liu, K. Li, Q. Su, and H. Yu, "A novel planar piezoelectric actuator with nano-positioning ability operating in bending-bending hybrid modes", *Ceram. Int.,* vol. 44, pp. S164-S167, 2018.
[http://dx.doi.org/10.1016/j.ceramint.2018.08.123]

[25]   S.K. Ghosh, K. Mallick, B. Tiwari, E. Sinha, and S.K. Rout, "Relaxor-ferroelectric BaLnZT (Ln = La, Nd, Sm, Eu, and Sc) ceramics for actuator and energy storage application", *Mater. Res. Express,* vol. 5, no. 1, p. 015509, 2018.
[http://dx.doi.org/10.1088/2053-1591/aaa4ee]

[26]   V. Pommier-Budinger, M. Budinger, P. Rouset, F. Dezitter, F. Huet, M. Wetterwald, and E. Bonaccurso, "Electromechanical Resonant Ice Protection Systems: Initiation of Fractures with Piezoelectric Actuators", *AIAA J.,* vol. 56, pp. 1-12, 2018.
[http://dx.doi.org/10.2514/1.J056662]

[27]   Q. Wang, and F. Li, "Large actuation strain over 0.3% in periodically orthogonal poled BaTiO$_3$ ceramics and multilayer actuators *via* reversible domain switching", *J. Phys. D Appl. Phys.,* vol. 51, no. 25, p. 255301, 2018.
[http://dx.doi.org/10.1088/1361-6463/aac3f2]

[28]   D. Bochenek, P. Niemiec, R. Skulski, and M. Adamczyk-Habrajska, "Electrophysical properties of a multicomponent PZT-type ceramics for actuator applications", *J. Phys. Chem. Solids,* vol. 133, pp. 128-134, 2019.
[http://dx.doi.org/10.1016/j.jpcs.2019.05.015]

[29]   X. Tian, B. Zhang, Y. Liu, S. Chen, and H. Yu, "A novel U-shaped stepping linear piezoelectric actuator with two driving feet and low motion coupling: Design, modeling and experiments", *Mech. Syst. Signal Process.,* vol. 124, pp. 679-695, 2019.
[http://dx.doi.org/10.1016/j.ymssp.2019.02.019]

[30]   X-Y. Kang, Z-H. Zhao, Y-K. Lv, and Y. Dai, "BNT-based multi-layer ceramic actuator with enhanced temperature stability", *J. Alloys Compd.,* vol. 771, pp. 541-546, 2019.
[http://dx.doi.org/10.1016/j.jallcom.2018.08.311]

[31]   M. Moretti, E.C.N. Silva, and J.N. Reddy, "Topology optimization of flextensional piezoelectric actuators with active control law", *Smart Mater. Struct.,* p. 28, 2019.
[http://dx.doi.org/10.1088/1361-665X/aafd56]

CHAPTER 2

Advanced Ceramics for Supercapacitors

N. Suresh Kumar[1,*], R. Padma Suvarna[1], K. Chandra Babu Naidu[2,*] and **Rajender Boddula[3]**

[1] *Department of Physics, JNTUA, Anantapuramu-515002, A.P, India*

[2] *Dept. of Physics, GITAM Deemed to be University, Bangalore-562163, Karnataka, India*

[3] *CAS Key Laboratory of Nano system and Hierarchical Fabrication, National Center for Nanoscience and Technology, Beijing 100190, PR China*

Abstract: In recent years, research is going on to introduce new types of electrode materials to satisfy the increasing demands for efficient energy storage devices. Among various energy storage devices, the supercapacitors gained a lot of attention owing to their long cyclic life, environment-friendly nature and high-power density. In this view, global research has been reported to this rapid development of fundamental and applied aspects of supercapacitors. In this chapter, we have given a brief description of different ceramic electrode materials along with their fabrication techniques for supercapacitor applications.

Keywords: Energy Density, Hydrothermal Technique, Molten Salt, Solution Precursor Flame Spray, Specific Capacitance, Supercapacitors.

2.1. INTRODUCTION

Etymologically, the term ceramic has been invented from two Greek words "keramos", which means "Potter's clay and "keramikos" means "clay products". Ceramic material is a "refractory, inorganic and non-metallic material which is often crystalline" or "the material which is formed by the use of heat". The advances in manufacturing techniques and the challenges posed by new applications led to the development of advanced ceramics. The capability of operating at temperatures greater than metals makes the ceramics a potential candidate for many structural applications like engine components, valves, bearings, *etc.* Ceramics having a low coefficient of thermal expansion and high thermal conductivity find applications in the field of electronics like capacitors,

* Corresponding authors Dr. K. Chandra Babu Naidu & Dr. N. Suresh Kumar: GITAM Deemed To Be University-Bangalore Campus, Bangalore-562163, Karnataka, India; E-mails: chandrababu954@gmail.com & sureshmsc6@gmail.com

superconductors, magnets and transducers. Modern engineered ceramics are known for their advantages like more strength, high operating temperatures, improved toughness, high resistance to melting, bending, stretching, corrosion, heat, physical stability, hardness, chemical inertness, biocompatibility, superior electrical properties, suitability in mass production and tailorable properties makes them one of the most versatile materials. When decreased density and larger melting points are required to enhance the efficiency and speed of the process, advanced ceramics are taking the place of metals.

2.2. ADVANCED CERAMICS

The bond between ceramic particles differentiates the advanced/engineering ceramics from conventional ones. In advanced ceramics, the particles are combined at grain-boundaries by means of similar energy equilibrium mechanism, whereas in conventional ceramics, particles are bonded by a weak interlocking or mechanical linking, because the impurities that are present will restrict the self-bonding of the particles. Based on their mechanism, the advanced ceramics are classified as metal oxide ceramics, glass ceramics, carbides and nitrides.

2.2.1. Metal Oxide Ceramics

Some of the metals are also used as the main constituent of ceramics called metal oxide ceramics. Alumina, beryllia and zirconia are examples of metal oxide ceramics which are used in advanced ceramics in their pure form. The features of the above said ceramics are discussed below.

Salient Features of Alumina

- It can be mixed with chromium/oxides of silicon/magnesium/calcium and can be used at 3500°F as long as they are not subjected to thermal shock and highly corrosive atmospheres.
- In general, alumina strength decreases beyond 3700°F.
- Up to 1500°F, they have good creep resistance.
- They are vulnerable to corrosion from steam, sodium and strong acids.

Salient Features of Beryllia

- Ceramic materials made up of Beryllia are expensive and excellent heat dissipators and electrical insulators as well.
- Possess extremely good thermal shock resistance, low coefficient of thermal expansion and high thermal conductivity.

Salient Features of Zirconia

- Of all ceramics, Zirconia is having the highest melting point which is about 4000°F.
- Strongest and toughest ceramics are made from transformation-toughened zirconia ceramics like zirconia strengthened alumina (ZTA), Y-TZP (yttria stabilized tetragonal zirconia polycrtystals) *etc.*
- Further, Y-TZP ceramics are used in pump and controller mechanisms where corrosion and wear resistance are required.
- ZTA ceramics possess low density, improved thermal shock resistance and economical than Y-TZP. These ceramics are used in transportation equipment.

2.2.2. Glass Ceramics

These types of ceramics are obtained from molten glass. On subjecting to heat, this molten glass will be crystallized. These ceramics find applications in cooking utensils, table ware, smoothe cooktops *etc.* By controlling the crystalline structure in the host glass matrix, properties can be altered. Lithium-aluminium-silicate (LAS or beta beta spondumene), magnetism-aluminium-silicate (MAS) and alumina-silicate (AS) are the mostly used glass ceramics because at higher temperatures, these three ceramics are stable, possess near-zero coefficients of thermal expansion and also oppose oxidation.

Salient Features of LAS

- No appreciable thermal expansion up to 800°C.
- Low thermal expansion due to high silica content.
- Strength also decreases due to silica.
- Attacked by sulfur and sodium.

Salient Features of MAS

- Stronger and corrosion resistant.
- MAS when mixed with aluminum titanate exhibits good corrosion resistance upto 2000°F.

Salient Features of AS

- Obtained by discharging lithium out of lithium aluminum silicate particles.
- It is corrosion resistant and strong.
- Utilized in turbine engines.

Another ceramic which is as stronger as alumina is machinable glass ceramic technically called as Macor. It possesses various electrical and high temperature properties and also can be machined with conventional tools.

2.2.3. Carbines and Nitrides

Boron carbine and nitride, silicon carbide and nitride, and aluminum nitride are the commonly used advanced engineering ceramics. Boron carbide has the following features, very high hardness and low density, excellent abrasion resistance and low strength at high temperatures. These ceramics find applications in light weight bullet proof armor plates and in pressure-blasting nozzles. The drawback with these ceramics is low strength at elevated temperatures and very expensive. Silicon nitride and silicon carbide are considered as high temperature, high strength "superstars" of modern ceramic materials because these ceramics are the sturdiest structural ceramics preferred for oxidation-resistant devices at high-temperatures. In this chapter, we have focussed on the role of ceramics in advancement of supercapacitors.

2.3. SUPERCAPACITORS

In the advancement of human life, energy plays an important role. Moreover, the world ecology and economy are affected by the utilization of fossil fuels owing to produce energy and to consume the same. Hence, the demand has been increased to develop environment, ecofriendly and renewable energy storage devices. The energy storage devices such as fuel cells, batteries and supercapacitors work on the principle of conversion of electrochemical energy. Among different storage devices, supercapacitors gained much attention due to high life cycle, large specific capacity and maintenance free [1 - 4]. In addition, these supercapacitors being flexible, light weight and small can also be utilized as a power supplier in portable devices such as digital cameras, mobile phones *etc*. Further, in heavy vehicles, the supercapacitors offer high power density which includes regaining of energy during the process of slow-down of the vehicle [5]. By using glass vessel with metal, foils Leyden jar was introduced, which is considered as the beginning of capacitor technology. In this Leyden jar, the jar acts as a dielectric and the foil of metal acts as electrodes. During the process of charging, positive and negative charges are accumulated on the electrodes and during discharging the two electrodes are connected by a metal wire [6]. In this chapter, we focussed on the ceramic based electrode materials for supercapacitor applications. The Table **2.1** represents some recently introduced electrode materials along with their specific capacitance, energy density, power density and cyclic stability.

Table 2.1. Ceramic electrode materials for supercapacitor applications.

S.No.	Material	Synthesis Technique	Specific Capacitance	Energy Density	Power Density	Retention of the Capacity	Ref.
1.	$Ni_2CoS_{4@}$ g-C_3N_4		2210 F/g at 1 A/g			85% retention after 4000 cycles	[37]
2.	Co (OH)$_2$	hydrothermal reaction	1066 F/g at 2 A/g	20.05 W h/Kg	1.22 W h/Kg	93% retention even after 10000 cycles	[30]
3.	$NiCo_2O_4$		1363 F/g at 2 A/g			82% retention at 20 A/g current density	[31]
4.	$NiCo_2S_4$/NCF	growth of precursor materials followed by sulfidation process	877 F/g at 20 A/g				[7]
5.	MnO_2/SA-ECNFs		141 F/g	12.5 Wh/kg			[32]
6.	$NiCo_2O_4$	Solution Precursor Flame Spray	902 F/g at 1 A/g			89.2% retention after 2500 cycles	[34]
7.	$CaCu_3Ti_4O_{12}$	molten salt technique	1185 mF/cm^2 areal capacitance at 10 mV/s			94% retention after 2500 cycles	[36]
8.	Fe doped $NiMnO_3$	microwave assisted hydrothermal technique	732.7 F/g at 1 A/g			78.2% retention after 10000 cycles	[21]
9.	CC/CWNA@Ni@ $CoNi_2S_4$	electrodeposition	3163 F/g at 1 A/g	53.8 Wh/kg	801 W/kg	90.1% retention after 10000 cycles	[38]
10.	$NiCo_2O_4$	co-electrodeposition	450 F/g at 20 A/g				[32]
11.	Ni-doped $MnCO_3$	hydrothermal technique	583.5 F/g 1 A/g	24.1 Wh/kg	0.74 kW/kg	84.78% retention after 2000 cycles	[20]

(Table 2.1) cont.....

S.No.	Material	Synthesis Technique	Specific Capacitance	Energy Density	Power Density	Retention of the Capacity	Ref.
12.	$Co_3O_4/MnCo_2O_{4.5}$		636 F/g at 1 A/g			86% retention after 5000 cycles	[35]

2.4. ADVANCED CERAMICS FOR SUPERCAPACITORS

A new type of electrode materials with suitable architecture is strongly required for enlightening the energy density of supercapacitors. Owing to high electrochemical activity and high specific capacity, binary metal sulfides gain much attention as electrode materials for high performance energy storage devices compared to mono-metal sulfides. In this connection, Shen *et al.* [7], demonstrated the design and fabrication of $NiCo_2S_4$ nano-sheets supported on NCF (Nitrogen doped Carbon Foams) as an electrode material for high performance supercapacitors. They reported that the synthesis of electrode materials involves two steps like the growth of Ni-Co precursors followed by $NiCo_2S_4$ conversion *via* sulfidation process. Furthermore, due to availability of rapid ways for electron and ion transportation in $NiCo_2S_4$/NCF electrode material, the $NiCo_2S_4$/NCF electrode material shows the high electrochemical performance, good cycling stability, and high specific capacitance of 877 F/g at a current density of 20 A/g. Also, utilizing the $NiCo_2S_4$/NCF as a cathode material and ordered mesoporous carbon-(OMC)/NCF as an anode material, they fabricated a device having simultaneous improvement in specific capacity, energy, power performance *etc.* So, the prepared electrode materials can be promising candidates for developing or fabricating the electrochemical devices for high energy storage applications. In KOH electrolyte, the electrochemical properties of Ti_3C_2 supercapacitors was investigated by Gao *et al.* [8], by synthesizing 2D (two dimensional) carbide Ti_3C_2 through exfoliating Ti_3AlC_2 in the solution of HF. The studies on electrochemical properties revealed that the Ti_3C_2 material has high SSA (specific surface area) about 22.35 m^2/g. In addition to this, the prepared material shows high energy storage capability and volumetric capacity of 119.79 F/cm^3 at 2.5 A/g current density. Further, they reported that the inclusion of carbon black to the prepared materials greatly enhances the electrochemical properties of the supercapacitors based on Ti_3C_2 materials. Hence, the 2D (two dimensional) carbide Ti_3C_2 materials can be served as potential electrode materials for supercapacitors with high storage and cycling ability. Lee and Koh [9], prepared $CaCu_3Ti_4O_{12}$ (CCTO) ceramic materials by mixed oxide technique and also sintered at different timings (30 min-12 h) at a temperature of 1125°C for supercapacitor applications. From the SEM analysis, it is clear that the mean grain size of CCTO increases with increasing sintering time. For instance, for the sintering at 30 min., the average grain size is 1-2 µm but it is 30-100 µm for 2 h sintering time. Also, they observed that at a sintering time of 2 h, the CCTO exhibits grains with large size and agglomerated parts consist of small grains and grain boundaries. The prepared CCTO ceramics exhibit the cubic structure which is confirmed by X-ray diffraction analysis. Moreover, increased sintering time leads to considerable enhancement in permittivity and its reduction with increasing frequency. The resistances of grain and grain boundaries are reduced with increasing temperature from 30-120°C. Therefore, the activation energies and resistivity of grain and grain boundaries are depending on sintering time.

Nowadays, the supercapacitors have been explored intensively [10 - 13]. Nevertheless, it is difficult to further enhance the supercapacitor performance to meet the increasing demand of the energy requirements. In the evolution of high-performance SCs, the researchers are focused on preparing modified and advanced electrode materials with outstanding electrochemical properties. Further, due to their high-power density compared to electronic double layer capacitive materials, the transition metal oxides (TMO) usually having multiple oxidation states such as graphene [14, 15], active carbon [16, 17], carbon nanotubes [18, 19] *etc.*, are normally used as the electrode materials for supercapacitors. In this view, the manganese carbonate ($MnCO_3$) gained much attraction as an electrode material for the next generation supercapacitor applications owing to its natural abundancy, low toxicity, high electrochemical cyclic life, high theoretical capacitance *etc.* Furthermore, the synthesis techniques also played a key role on the performance of the electrode materials. Generally, some techniques are involved in doping and morphological adjustment of the homogeneous elements which are commonly used to enhance the performance of the electrode materials. In this sequence, Zhao *et al.* [20], prepared Ni-doped $MnCO_3$ ceramic materials *via* single step hydrothermal technique. The prepared materials exhibit 3D flower like structure, large surface area especially 3D self-assembled $Ni_{0.2}Mn_{0.8}CO_3$ nano flowers exhibiting good electrochemical properties, with area capacitance of 583.5 F/g at a current density of 1 A/g and 84.78% capacitance retention even after 2000 cycles at 2 A/g. Besides, the ASC assembled with activated carbon as an anode and the prepared material as anode displays high cycling stability as well as large energy density of 24.1 Wh/kg at 0.74 kW/kg power density. Hence, the prepared material will be a prominent electrode material for high performance supercapacitors.

Looking for the new electrode materials, Qiao *et al.* [21], synthesized Fe doped $NiMnO_3$ ceramic materials *via* microwave assisted hydrothermal technique for the supercapacitor applications. The prepared materials can serve as a new kind of electrode material for the supercapacitors. From the different characterizations, they reported that the structure as well as the morphology of $NiMnO_3$ is changed by Fe doping. Further, Fe-doped $NiMnO_3$ electrode material exhibits excellent conductivity with high surface area. Moreover, the electrode materials show considerable electrochemical performance which is confirmed *via* electrochemical impedance spectroscopy (EIS), galvanostatic charge/discharge (GCD) and cyclic voltammetry (CV) studies. At 1 A/g current density, the prepared electrode material shows the high specific capacitance of 732.7 F/g and the same material at 3 A/g current density exhibits 78.3% capacitance retention even after 10000 cycles. This long cyclic life, high specific capacitance and large surface area suggested that the Fe-doped $NiMnO_3$ ceramic can be a promising electrode material for the next generation supercapacitors. Yuan *et al.* [22], proposed a simple two step technique for big scale of growth of mesoporous $NiCo_2O_4$ nano sheets on conductive Ni-foam with strong bond as an electrode material for high performance supercapacitors. The fabrication encompasses the co-electrodeposition of a (Ni, Co) bimetallic hydroxide precursor on a nickel foam support and succeeding thermal-transformation to spinel-mesoporous nickel cobaltite. They reported that the prepared electrode materials exhibit the thickness in nano range of the order 5 nm. In addition to this, the prepared electrode material shows good structural stability, high electroactive surface area and fast ion & electron transport. The specific capacitance around 1450 F/g at a high current density of 20 A/g and good cycling ability was obtained. This provides a new technique for constructing advanced electrode materials with large specific capacitance for supercapacitor applications.

Numerous researchers have introduced new kinds of liquid electrolytes for the advancement of supercapacitors which play a vital role in various applications like power backup, industrial energy management system, electric vehicle *etc* [23 - 25]. But, one of the major disadvantages of the liquid electrolytes in supercapacitors is the potential safety problems [26]. Thus, the replacement of liquid electrolyte with solid state electrolyte is a direct and effective solution for this. In addition, these solid electrolytes have some advantages rather than liquid electrolytes such as high ionic conductivity and wide range of operation temperature [27, 28]. Hence, these solid-state electrolytes play a key role in the advancement of next generation supercapacitors. In this connection, with the help of conventional solid-state reaction technique, Gu *et al.* [29], prepared the LLTO ($Li_{0.5}La_{0.5}TiO_3$) ceramics for supercapacitor applications. They observed that the prepared materials exhibit tetragonal structure. Further, the SEM and EDAX pictures revealed that there exist clear interfaces between the prepared ceramics and metal (Ag, Au and Pt) electrodes. Figure 2 of reference [29] represents the SEM and EDAX images of the $Li_{0.5}La_{0.5}TiO_3$ ceramics and Ag, Au and Pt electrodes. In addition, for diverse metal electrodes at relatively low frequency, the prepared LLTO ceramic materials exhibit colossal nominal permittivity.

Colossal permittivity is mainly due to grain boundary interfacial polarization which is confirmed from AC impedance spectra. The value of leakage current for LLTO/Ag electrode is less than that of LLTO/Pt and LLTO/Au electrodes. Besides, the presence of electrochemical reactions between LLTO/Au and LLTO/Pt electrodes is confirmed from the redox current peaks in I-V curves. From these results, it is clear that for next generation supercapacitors, the LLTO ceramics can be served as potential electrode materials.

Due to large theoretical capacitance which is attained by charge storing mechanism, the cobalt hydroxides gained much attention for the electrode materials for supercapacitors. Further, $Co(OH)_2$ in crystalline phase is classified into two polymorphs namely α and β polymorphs. The α-phase contains of positively charged $Co(OH)_{2-x}$ which exhibits hydrotalcite-like structure, whereas β-phase is compiled of brucite-like structure. For the applications of high-performance asymmetric supercapacitors, Jana *et al.* [30], synthesized $Co(OH)_2$ using IL (ionic-liquid) of $[BMim][BF_4]$ (1-butyl-3methylimidazolium tetrafluoroborate) through hydrothermal reaction technique. Herein, the crystalline phases α-$Co(OH)_2$ and β-$Co(OH)_2$ cobalt hydroxide were controlled using ionic liquid. Here, α-$Co (OH)_2$ consists of both tetrahedral and octahedral Co sites while β-$Co(OH)_2$ consists only octahedral Co-sites. Further, the IL plays a vital role as a co-solvent and prototype to control the morphology of cobalt hydroxide. From the morphological analysis, it is clear that the α-$Co(OH)_2$ exhibits flake like structure and β-$Co(OH)_2$ displays nanorod like structure. The interlayer spacing of α-$Co(OH)_2$ is of the order of 8.24 Å in contrast to β-$Co(OH)_2$ which shows the interlayer spacing of 4.63 Å. The addition of hydroxide anions to active α-$Co(OH)_2$ is prevented by the precursor anions, which exhibit the 613 F/g specific capacitance at 2 A/g current density, whereas β-$Co(OH)_2$ exhibits the large specific capacitance of 1066 F/g at 2 A/g. In addition, the asymmetric supercapacitor with merged β-$Co(OH)_2$ and RGO electrodes delivers high energy-density of 20.05 Wh/Kg at the power density of 1.22 Wh/Kg. Furthermore, the asymmetric supercapacitor maintains the energy density moderately of the order of 19.22 Wh/Kg at large power density of 13.40 Wh/Kg and 93% retention even after 10000 cycles. Hence, these outcomes indicate that this new approach of synthesizing cobalt-based materials for supercapacitor applications exhibits high specific capacity with long cyclic life.

Generally, several steps are involved in the preparation of carbon based $NiCo_2O_4$ electrode materials for battery applications which also requires very high temperatures. To overcome this, Geiuli *et al.* [31], proposed a single step strategy for the fabrication of MWCNTs (Multi-walled Carbon Nanotubes) decorated with NiO_x and nickel foam decorated with $NiCo_2O_4$ nanoparticles for developing a stable supercapacitor with large specific capacity. Here, the proposed technique is mainly based on the electrophoretic co-deposition of $NiCo_2O_4$ nanoparticles and carboxylated carbon nanotubes, which are charged positively duo to Ni^{2+} ions. The electrochemical performance can be improved through decorating the nanomaterials with the species of NiO_x. In addition, the electron transfer takes place effectively due to well dispersed $NiCo_2O_4$ nanoparticles in the CNT porous matrix. Besides, they also provide good cycling stability and large surface area. The electrochemical studies revealed that the prepared nano-composite exhibits large specific capacity of 1363 F/g at a current density of 2 A/g. However, it shows 82% retention at 20 A/g current density. With these outcomes, and using the one step strategy, we can prepare electrode materials with high electrochemical performance for supercapacitor applications. Large pseudocapacitance can be introduced by the presence of transition metal oxides in an EDLC (electric double layer capacitor) which improves the storing capacity of the capacitor and also, they possess large dielectric constant. Liu *et al.* [32], proposed a new technique called self-sustained bi-functional configuration for metal oxide supercapacitor in which the supercapacitor shows good energy storage performance which is achieved by eradicating traditional separator of a metal oxide film in supercapacitor. They reported that instead of traditional separator in the proposed supercapacitor, α-MnO_2 is uniformly electrodepositing on allied SA-ECNFs (super aligned electro spun carbon nanofibers). The viability of α-MnO_2 layer works as pseudocapacitive material and also works as separator, Figure 2 of reference [32] represents the schematic diagram of the electrodeposition of MnO_2 cathode in the proposed supercapacitor. The electrochemical studies revealed that MnO_2/SA-ECNFs electrode material with 40 μA electrodeposition show high specific energy density of 12.5 Wh/kg and specific capacitance of 141 F/g. Therefore, the proposed technique provides the pseudocapacitance energy storage replacing the separator with metal oxide/carbonaceous nanomaterial which also shortens the manufacturing process. Using the conventional solid-state technique, Lu *et al.* [33], prepared the $Li_{0.33}Sr_xLa_{0.56-2/3x}TiO_3$ (LSLT) ceramic materials for the supercapacitor applications. They reported that the prepared material is undergone various characterizations such as XRD (X-ray diffraction), SEM (scanning electron microscope), XPS (X-ray photoelectron spectroscopy) and electrochemical properties. From the outcomes, it is clear that the LSLT ceramic materials with 0.25% concentration exhibit high electrical conductivity, which is mainly due to the absence of impurities in the prepared materials, although the occupation of Sr is in the sites of Li and La. The occupation of Sr in Li and La sites leads to the breakage of bond between Li and La. The electrochemical studies revealed that the prepared materials exhibit the specific capacitance of 0.48 mF/g. Therefore, the LSLT ceramic material can be a potential candidate for the solid-state ionic supercapacitor.

Owing to highly active surface area and remarkable electrochemical properties the hallow nano/microstructures gained much attention for supercapacitor applications. In this context, Ma *et al.* [34], proposed a novel one-Step Solution Precursor Thermal Spray (SPTS) technology also called as Solution Precursor Flame Spray (SPFS) to synthesize and depositing the hallow microspheres on other metal oxides for the applications of supercapacitors. Figure 1b of reference [34], represents the schematic diagram of SPFS. They reported that by employing the above said technique for the first time, they prepared and studied the electrochemical properties of oxygen-vacancies $NiCo_2O_4$ based films with hollow microspheres deposited on Ni-substrate. The presence of $NiCo_2O_4$ is confirmed by the XRD, XPS and Raman spectra. Especially, the cyclic-voltammetry (CV) test revealed that at a current density of 1 A/g, the SPTS-deposited films exhibited a large specific-capacitance of 373 F/g which is almost 20 times better than Solution Precursor Plasma Spray (SPPS) deposited films. The schematic illustration of SPPS is shown in Figure 1a of reference [34]. In addition, the same films show the highest specific capacitance of 902 F/g. Further, after 2500 cycles, the prepared oxygen vacant films exhibit almost 89.2% retention at large current density of 20 A/g. Hence, the more oxygen vacant and hallow microsphere deposited films show the superior performance than the bulk which consist of less oxygen vacancies. There exist various monometallic oxides among all cobalt oxides, manganese oxides and bimetallic oxides which exhibit good electrochemical properties owing to high capacity, good conductivity and synergistic effect. Recently, Wang *et al.* [35], proposed a new core shell polyhedron $Co_3O_4/MnCo_2O_{4.5}$ as an electrode material for supercapacitor app-lications. They reported that to prepare the electrode material $Co_3O_4/ MnCo_2O_{4.5}$ with core shell structure, they combined Mn with ZIF-67 to form multi-shell structure followed by thermal treatment. Owing to their unique properties, the prepared material exhibits good specific capacity of 636 F/g at 1 A/g and good cycling life *i.e.*, after 5000 charge discharge cycles, it shows 86% of capacitance retention. Hence, it might be a prominent candidate for next generation supercapacitors. Maity *et al.* [36], investigated the electrochemical properties of $CaCu_3Ti_4O_{12}$ (CCTO) nanostructures which are synthesized *via* molten salt technique. The prepared ceramic nanocubes exhibit pure and single crystalline phases as confirmed from XRD and HRTEM analysis. Furthermore, the EIS, GCD and CV studies revealed that at a scan rate of 10 mV/s, the prepared nanostructures exhibit maximum areal capacitance of 1185 mF/cm^2 in KOH solution. Besides, the solid-state symmetric device fabricated by using the prepared nanocubes shows an outstanding cycling stability (94.8% retention) after 3000 cycles at a 10 mA/cm^2 current density. From these outcomes, it is clear that CCTO nanostructures can serve as a prominent electrode material for the next generation supercapacitors.

For introducing new kinds of the materials for the supercapacitor applications, with the help of hydrothermal technique, Ensafi *et al.* [37], coated Ni_2CoS_4 nanoparticles on the surface of g-C_3N_4 (graphitic carbon nitride). The prepared nanocomposites were undergone different characterization techniques. From the outcomes, they reported that $Ni_2CoS_4@g$-C_3N_4 nanocomposites exhibit superior electrochemical properties such as at 1 A/g discharge current, the composite shows the charge storage capacity of 2210 F/g or 246 mAh/g, and large cyclic-stability (even after 4000 charge-discharge cycles). The $Ni_2CoS_4@g$-C_3N_4 nanocomposites show 85% retention, 99% of columbic efficiency at 40 A/g discharge current. From all these values, the $Ni_2CoS_4@g$-C_3N_4 nanocomposites can be the potential electrode material for the advanced energy storage devices owing to large cyclic stability and columbic efficiency. In addition to this, owing to the high theoretical capacitance, the $CoNi_2S_4$ (Nickel cobalt sulfide) based materials have gained much attention as an electrode material for supercapacitors. But, due to low electrochemical stability and electrical conductivity, their applications are limited. To overcome this, using carbon cloth as secondary substrate, Wang *et al.* [38], fabricated MOF-CNWAs (metal organic framework derived carbon nano wall arrays) which provide more sites for electrodeposition of Nickel cobalt sulfide. From the outcome of electrochemical studies, they reported that the prepared electrode ($CC/CWNA@Ni@ CoNi_2S_4$) exhibits the outstanding specific capacitance of about 3163 F/g at a current density of 1 A/g or 2825 F/g at 5 mV/s. Besides, the prepared electrode materials show 53.8 Wh/kg energy density at 801 W/kg power density and even after 10000 cycles, it shows the 90.1% retention of specific capacitance. Hence, this long cyclic stability and high specific capacitance indicates that the $CC/CWNA@Ni@CoNi_2S_4$ electrode material is auspicious material for next generation supercapacitors. Hence, research is going on to improve the efficiency of the electrode materials for supercapacitor applications. In this chapter, we have discussed briefly about some of the advanced ceramic materials for next generation supercapacitors.

CONCLUSIONS

This chapter presents briefly an introduction of ceramics, different types of ceramic materials and role of ceramic materials in the evaluation of supercapacitors. Further, we focussed on different advanced ceramic electrode materials such as $Ni_2CoS_4@g$-C_3N_4, MnO_2/ SA-ECNFs, $CC/CWNA@Ni@CoNi_2S_4$, $Co_3O_4/MnCo_2O_{4.5}$ *etc.*, for the supercapacitor applications. In addition to this, some fabrication techniques like hydrothermal technique, molten salt technique, solution precursor flame spray *etc.*, are discussed as well.

CONSENT FOR PUBLICATION

Not applicable.

CONFLICT OF INTEREST

The authors confirm that this chapter content has no conflict of interest.

ACKNOWLEDGEMENTS

The authors express thankfulness to Dr. P. Sreeramulu, Assistant Professor (English), GITAM, Bangalore for providing English language editing services to this manuscript.

REFERENCES

[1]   L.L. Zhang, and X.S. Zhao, "Carbon-based materials as supercapacitor electrodes", *Chem. Soc. Rev.,* vol. 38, no. 9, pp. 2520-2531, 2009.
[http://dx.doi.org/10.1039/b813846j] [PMID: 19690733]

[2]   A. Balducci, R. Dugas, P. Taberna, P. Simon, D. Plee, M. Mastragostino, and S. Passerini, "High temperature carbon-carbon supercapacitor using ionic liquid as electrolyte", *J. Power Sources,* vol. 165, pp. 922-927, 2007.
[http://dx.doi.org/10.1016/j.jpowsour.2006.12.048]

[3]   C. Largeot, C. Portet, J. Chmiola, P.L. Taberna, Y. Gogotsi, and P. Simon, "Relation between the ion size and pore size for an electric double-layer capacitor", *J. Am. Chem. Soc.,* vol. 130, no. 9, pp. 2730-2731, 2008.
[http://dx.doi.org/10.1021/ja7106178] [PMID: 18257568]

[4]   S. Kandalkar, D. Dhawale, C. Kim, and C. Lokhande, "Chemical synthesis of cobalt oxide thin film electrode for supercapacitor application", *Synth. Met.,* vol. 160, pp. 1299-1302, 2010.
[http://dx.doi.org/10.1016/j.synthmet.2010.04.003]

[5]   M. Winter, and R.J. Brodd, "What are batteries, fuel cells, and supercapacitors?", *Chem. Rev.,* vol. 104, no. 10, pp. 4245-4269, 2004.
[http://dx.doi.org/10.1021/cr020730k] [PMID: 15669155]

[6]   B.E. Conway, "Transition from "Supercapacitor" to "Battery" behavior in electrochemical energy storage", *J. Electrochem. Soc.,* vol. 138, pp. 1539-1548, 1991.
[http://dx.doi.org/10.1149/1.2085829]

[7]   L. Shen, J. Wang, G. Xu, H. Li, H. Dou, and X. Zhang, "NiCo2S4 Nanosheets Grown on Nitrogen-Doped Carbon Foams as an Advanced Electrode for Supercapacitors", *Adv. Energy Mater.,* vol. 5, no. 3, 2014.1400977
[http://dx.doi.org/10.1002/aenm.201400977]

[8]   Y. Gao, L. Wang, Z. Li, Y. Zhang, B. Xing, C. Zhang, and A. Zhou, "Electrochemical performance of Ti3C2 supercapacitors in KOH electrolyte", *J Adv Ceramics,* vol. 4, no. 2, pp. 130-134, 2015.
[http://dx.doi.org/10.1007/s40145-015-0143-3]

[9]   J-W. Lee, and J-H. Koh, "Grain size effects on the dielectric properties of CaCu3Ti4O12 ceramics for supercapacitor applications", *Ceram. Int.,* vol. 41, no. 9, pp. 10442-10447, 2015.
[http://dx.doi.org/10.1016/j.ceramint.2015.04.109]

[10]   Z. Wang, and J. Cheng, "Zhou, J. Zhang J., Huang, H. J. Yang, Y. Li, B. Wang, All-climate aqueous fiber-shaped supercapacitors with record areal energy density and high safety", *Nano Energy,* vol. 50,

pp. 106-117, 2018.
[http://dx.doi.org/10.1016/j.nanoen.2018.05.029]

[11]   Z. Hou, F. Tian, Y. Gao, W. Wu, L. Yang, X. Jiao, and K. Huang, "Nickel cobalt hydroxide/reduced graphene oxide/carbon nanotubes for high performance aqueous asymmetric supercapacitors", *J. Alloys Compd.,* vol. 753, pp. 525-531, 2018.
[http://dx.doi.org/10.1016/j.jallcom.2018.04.245]

[12]   M. Yu, Y. Lu, H. Zheng, and X. Lu, "New insights into the operating voltage of aqueous supercapacitors", *Chemistry,* vol. 24, no. 15, pp. 3639-3649, 2018.
[http://dx.doi.org/10.1002/chem.201704420] [PMID: 29024125]

[13]   C. An, W. Li, M. Wang, Q. Deng, and Y. Wang, "High energy density aqueous asymmetric supercapacitors based on MnO_2@C branch dendrite nanoarchitectures", *Electrochim. Acta,* vol. 238, pp. 603-610, 2018.
[http://dx.doi.org/10.1016/j.electacta.2018.07.006]

[14]   Y. Bi, A. Nautiyal, H. Zhang, H. Yan, J. Luo, and X. Zhang, "Facile and ultrafast solid-state microwave approach to MnO2-NW@Graphite nanocomposites for supercapacitors", *Ceram. Int.,* vol. 44, pp. 5402-5410, 2018.
[http://dx.doi.org/10.1016/j.ceramint.2017.12.169]

[15]   Q. Zhang, Y. Wang, B. Zhang, K. Zhao, P. He, and B. Huang, "3D super elastic graphene aerogel-nanosheet hybrid hierarchical nanostructures as high-performance supercapacitor electrodes", *Carbon,* vol. 127, pp. 449-458, 2018.
[http://dx.doi.org/10.1016/j.carbon.2017.11.037]

[16]   X. Zhu, S. Yu, K. Xu, Y. Zhang, L. Zhang, G. Lou, Y. Wu, E. Zhu, H. Chen, Z. Shen, B. Bao, and S. Fu, "Sustainable activated carbons from dead ginkgo leaves for supercapacitor electrode active materials", *Chem. Eng. Sci.,* vol. 181, pp. 36-45, 2018.
[http://dx.doi.org/10.1016/j.ces.2018.02.004]

[17]   H. Guo, Z. Liu, H. Li, H. Wu, C. Zhang, J. Yang, and X. Chen, "Active carbon electrode fabricated *via* large-scale coating-transfer process for high-performance supercapacitor", *Appl. Phys., A Mater. Sci. Process.,* p. 123, 2017.
[http://dx.doi.org/10.1007/s00339-017-1069-0]

[18]   H. Li, F. Yue, C. Yang, P. Qiu, Q. Xue, P. Xu, and J. Wang, "Porous nanotubes derived from a metal-organic framework as high-performance supercapacitor electrodes", *Ceram. Int.,* vol. 42, pp. 3121-3129, 2016.
[http://dx.doi.org/10.1016/j.ceramint.2015.10.101]

[19]   R. Mohan, and R. Paulose, "An efficient electrochemical performance of Fe2O3/CNT nanocomposite coated dried Lagenaria siceraria shell electrode for electrochemical capacitor", *Ceram. Int.,* vol. 44, pp. 10990-10993, 2018.
[http://dx.doi.org/10.1016/j.ceramint.2018.03.154]

[20]   N. Zhao, H. Fan, M. Zhang, X. Ren, C. Wang, H. Peng, and X. Cao, "Facile preparation of Ni-doped MnCO3 materials with controlled morphology for high-performance supercapacitor electrodes", *Ceram. Int.,* vol. 45, no. 5, pp. 5266-5275, 2018.
[http://dx.doi.org/10.1016/j.ceramint.2018.11.224]

[21]   S. Qiao, N. Huang, J. Zhang, Y. Zhang, Y. Sun, and Z. Gao, "Microwave-assisted synthesis of Fe-doped NiMnO3 as electrode material for high-performance supercapacitors", *J. Solid State Electrochem.,* vol. 23, pp. 63-72, 2019.
[http://dx.doi.org/10.1007/s10008-018-4115-8]

[22]   H. Yuan, J. Li, L. Hou, X. Zhang, L. Shen, and X. Lou, "Ultrathin nanosheets supported on ni foam as advanced electrodes for supercapacitors", *Adv. Funct. Mater.,* vol. 22, pp. 4592-4597, 2012.
[http://dx.doi.org/10.1002/adfm.201200994]

[23]   J. Ren, L. Li, C. Chen, X. Chen, Z. Cai, L. Qiu, Y. Wang, X. Zhu, and H. Peng, "Twisting carbon

nanotube fibers for both wire-shaped micro-supercapacitor and micro-battery", *Adv. Mater.,* vol. 25, no. 8, pp. 1155-1159, 1224, 2013.
[http://dx.doi.org/10.1002/adma.201203445] [PMID: 23172722]

[24]   C. Meng, C. Liu, L. Chen, C. Hu, and S. Fan, "Highly flexible and all-solid-state paperlike polymer supercapacitors", *Nano Lett.,* vol. 10, no. 10, pp. 4025-4031, 2010.
[http://dx.doi.org/10.1021/nl1019672] [PMID: 20831255]

[25]   B.G. Choi, J. Hong, W.H. Hong, P.T. Hammond, and H. Park, "Facilitated ion transport in all-soli--state flexible supercapacitors", *ACS Nano,* vol. 5, no. 9, pp. 7205-7213, 2011.
[http://dx.doi.org/10.1021/nn202020w] [PMID: 21823578]

[26]   F. Meng, and Y. Ding, "Sub-micrometer-thick all-solid-state supercapacitors with high power and energy densities", *Adv. Mater.,* vol. 23, no. 35, pp. 4098-4102, 2011.
[http://dx.doi.org/10.1002/adma.201101678] [PMID: 21815221]

[27]   D.N. Futaba, K. Hata, T. Yamada, T. Hiraoka, Y. Hayamizu, Y. Kakudate, O. Tanaike, H. Hatori, M. Yumura, and S. Iijima, "Shape-engineerable and highly densely packed single-walled carbon nanotubes and their application as super-capacitor electrodes", *Nat. Mater.,* vol. 5, no. 12, pp. 987-994, 2006.
[http://dx.doi.org/10.1038/nmat1782] [PMID: 17128258]

[28]   M. Kaempgen, C.K. Chan, J. Ma, Y. Cui, and G. Gruner, "Printable thin film supercapacitors using single-walled carbon nanotubes", *Nano Lett.,* vol. 9, no. 5, pp. 1872-1876, 2009.
[http://dx.doi.org/10.1021/nl8038579] [PMID: 19348455]

[29]   Rui Gu, Kun Yu, Lingfeng Wu, Ruiping Ma, Hongchen Sun, Li Jin, Youlong Xu, Zhuo Xu, and Xiaoyong Wei, "Dielectric properties and I-V characteristics of Li0.5La0.5TiO3 solid electrolyte for ceramic supercapacitors", *Ceramics International,* vol. 45, no. 7, pp. 8243-8247, 2019.

[30]   M. Jana, P. Sivakumar, M. Kota, M.G. Jung, and H.S. Park, "Phase- and interlayer spacing-controlled cobalt hydroxides for high performance asymmetric supercapacitor applications", *J. Power Sources,* vol. 422, pp. 9-17, 2019.
[http://dx.doi.org/10.1016/j.jpowsour.2019.03.019]

[31]   O. Geuli, Q. Hao, and D. Mandler, "One-step fabrication of NiO_x-decorated carbon nanotubes-$NiCo_2O_4$ as an advanced electroactive composite for supercapacitors", *Electrochim. Acta,* vol. 318, pp. 51-60, 2019.
[http://dx.doi.org/10.1016/j.electacta.2019.06.044]

[32]   Y. Liu, Z. Zeng, R.K. Sharma, S. Gbewonyo, K. Allado, L. Zhang, and J. Wei, "A bi-functional configuration for a metal-oxide film supercapacitor", *J. Power Sources,* vol. 409, pp. 1-5, 2019.
[http://dx.doi.org/10.1016/j.jpowsour.2018.10.084]

[33]   D. Lu, J. Ma, J. Wu, Y. Yao, T. Tao, B. Liang, and S. Lu, "Preparation and electrochemical properties of $Li_{0.33}SrLa_{0.56-2/3}TiO_3$-based solid-state ionic supercapacitor", *Ceram. Int.,* vol. 45, no. 2, pp. 2584-2590, 2019.
[http://dx.doi.org/10.1016/j.ceramint.2018.10.192]

[34]   Y. Ma, Z. Yu, M. Liu, C. Song, X. Huang, M. Moliere, and H. Liao, "Deposition of binder-free oxygen-vacancies $NiCo_2O_4$ based films with hollow microspheres *via* solution precursor thermal spray for supercapacitors", *Ceram. Int.,* vol. 45, no. 8, pp. 10722-10732, 2019.
[http://dx.doi.org/10.1016/j.ceramint.2019.02.145]

[35]   Z. Wang, F. Wei, Y. Sui, J. Qi, Y. He, and Q. Meng, "A novel core-shell polyhedron $Co_3O_4/MnCo_2O_{4.5}$ as electrode materials for supercapacitors", *Ceram. Int.,* vol. 45, no. 9, pp. 12558-12562, 2019.
[http://dx.doi.org/10.1016/j.ceramint.2019.03.010]

[36]   S. Maity, M. Samanta, A. Sen, and K.K. Chattopadhyay, "Investigation of electrochemical performances of ceramic oxide $CaCu_3Ti_4O_{12}$ nanostructures", *J. Solid State Chem.,* vol. 269, pp. 600-607, 2019.
[http://dx.doi.org/10.1016/j.jssc.2018.10.016]

[37]   A.A. Ensafi, M.M. Abarghoui, and B. Rezaei, "Graphitic carbon nitride nanosheets coated with Ni_2CoS_4 nanoparticles as a high-rate electrode material for supercapacitor application", *Ceram. Int.,* vol. 45, no. 7, pp. 8518-8524, 2019.
[http://dx.doi.org/10.1016/j.ceramint.2019.01.165]

[38]   Y. Wang, J. Huang, Y. Xiao, Z. Peng, K. Yuan, L. Tan, and Y. Chen, "Hierarchical nickel cobalt sulfide nanosheet on MOF-derived carbon nano wall arrays with remarkable super capacitive performance", *Carbon,* vol. 147, pp. 146-153, 2019.

CHAPTER 3

Advanced Ceramics for Magnetocaloric Effect in Refrigerators

U. Naresh[1], N. Suresh Kumar[2], K. Chandra Babu Naidu[3,*] and D. Ravinder[4]

[1] *Department of Physics, BIT Institute of Technology, Hindupuram-515502, A.P, India*

[2] *Department of Physics, JNTUA, Anantapuramu-515002, A.P, India*

[3] *Dept. of Physics, GITAM Deemed to be University, Bangalore-562163, Karnataka, India*

[4] *Department of Physics, Osmania University, Hyderabad, 500007, Telangana, India*

Abstract: In this chapter, we have reviewed the advanced ceramics which are used for the refrigeration or cooling process. The refrigeration process is carried out using the magnetocaloric effect, *i.e.*, the variation of specific heat and thermal energy as a function of temperature and magnetic field. The ceramics of magnetite and rare earth alloys of magnetic composites are briefly discussed.

Keywords: Ceramics, Ferrites, MCE, Refrigeration.

3.1. INTRODUCTION

The magnetic refrigeration derived from the magnetocaloric effect played a vital role in viewing the environment-friendly technology. From the last decade, the research in the direction of the magnetocaloric based materials and their characterization techniques were improved proportionally. The major cause of the existing this technology is its terrific alternative technology rather than the used gas compression or expansion-based refrigeration technology. In this, a magnetic solid is used as the refrigerator agent rather than the sources of greenhouse effect in the refrigerator [1 - 5]. In the commonly used refrigeration process, the fluids such as isobutene, carbon dioxide have been used as a solution. This makes flammability and hence the researchers get interest in this field. The use of ceramic materials in refrigeration cooling technology has resulted in research. Magneto-caloric materials are the part of the magneto-ceramic material offering application for cooling at room temperature in the place of hazardous and traditional gas refrigerators. Mainly, in the last few decades, some of the theore-

---

* **Corresponding author Dr. K. Chandra Babu Naidu:** GITAM Deemed To Be University-Bangalore Campus, Bangalore-562163, Karnataka, India; E-mail: chandrababu954@gmail.com

tical studies show rod like magnetic materials which maximize a refrigeration device capacity [6]. In addition to this, the ceramic material possesses the improved mechanical and chemical stability with the organic and inorganic metals and polymers [7]. The magnetocaloric effect (MCE) is defined as the temperature changes from maximum to minimum and then it reaches to minimum upon application of external magnetic field on the magnetic material. It was theoretically explained by Warburg in 1881, later many of researchers continued studies on this effect, in 1949 the scientist Giauque gets Noble Prize in Chemistry as obtaining MCE effectively.

Generally, the MCE is a magneto-thermal phenomenon which is connected to coupling between lattice and external magnetic field. The MCE can be formulated based on the magnetic domain entropy (D_{SM}) and changes with temperature (Dt_{ad}) adiabatically, if the magnetic ceramics exposed to the external varying magnetic field [8]. The MCE is an aspect of some magnetic materials when the external magnetic field applied. Then, the heat up variation is taken place. Furthermore, the alignment of the atomic magnetic dipoles depends on the direction of the applied field and they become cool-down, once the applied field is removed. The material which exhibits this effect is named as MCMs (magnetocaloric materials). The materials can be characterized through the way of measuring temperature variation upon application of external magnetic field. The main parameters for the calculating MCE are the isothermal entropy variation Δs_{iso} and adiabatic temperature ΔT_{ad} of magneto caloric materials below external magnetic field variation at adiabatic conditions. The experiments are carried out through direct and indirect measurements to get MCE properties to obtaining AMR modeling and adiabatic temperature variation [9 - 11]. The direct method is temperature dependent technique in which the MCM is exposed to external field. Precisely, one can attain the adiabatic temperature change, whereas the second one depends on the magnetization and heat capacity measurements. In the indirect method, by studying the magnetizing curves which were obtained at temperature and applied field through isothermal procedure. Based on the thermodynamic model and first principle models, we can approach to evaluate adiabatic temperature variation of the system [12]. In particular, the thermodynamic model is commonly utilized in magnetic refrigeration to estimate the MCE in mathematical models of AMR-cycles [13 - 16]. By using the state of equations, we can identify the relation among the applied magnetic field, temperature and magnetization. However, the Weiss mean field theory (MFT) is commonly used as thermodynamic model for the evaluation of the magnetocaloric effect. Also, the MFT was applied to calculate the total entropy of the ferromagnetic material using the temperature variation with respect to external magnetic field [17]. Although in the first-principle models, it is found on an estimation of the interchange magnetic moments and coupling energies of magnetocaloric materials.

3.2. MAGNETOCALORIC MATERIALS

In following section, some of the crystalline materials are considered which the parts of the magnetocaloric materials. Recently, the magnetocaloric materials are mainly synthesized using the rare earth (RE) elements. Also, they define the parameters of MCE in rare earth free systems. Further, in the year 2000, Gschneidner & Pecharsky [10] studied the magnetocaloric nature of pure lanthanide elements. Nevertheless, one must indicate that in certain RE metals, there exists the robust improvement of MCE for small changes in field over ultrafine layers when their thickness is under the helix-period. Then, no magnetic order is formed, wherein the Ho-ultrathin film, the magnetocaloric curves demonstrated the shape like caret for traditional ferromagnetic materials [18, 19]. The Curie temperature of T_C=121 K, and T_{ad} was improved as a function of magnitude in regard to bulk samples (T_{ad} =1.3 K and 0.08 K for H =0.2 T for thin film and bulk samples, respectively) [20]. In the case of cubic phases (Fd3m space group), the rare earth atoms create a diamond unit cell while the remaining atoms are surrounded by the rare earth atoms establishing tetrahedral. In that lattice system, two magnetic sublattices come into view. They are: one is observed *via* magnetic rare earth atoms and other observed *via* traveling electron-magnetism owing to TMs. Herein, the $RECo_2$ composites, both magnetic sublattices are coupled for light RE-elements (for example Sm, Pr and Nd are ferro magnets) and coupled for anti-parallel weighty RE elements (for example Er, Ho, Gd, Dy and Tb are ferrimagnets).

Shen *et al.* [21], recently, reported a system of binary La-Fe phase-diagram which displays an immiscible-system; wherein no inter-metallic composites are developed. The mixing of minor quantity of Si/Al permits the development of ferromagnetic systems with FCC-$NaZn_{13}$-type structure comprising of 112 atoms per unit cells. These inter-metallic complexes show identical magnetic nature. Hu *et al.* [22], reported that for $LaFe_{11.4}Si_{1.6}$, Δs_{iso} is nearer to 20 Jkg^{-1}K^{-1} at 210 KJkg^{-1} for H=5T. It was attributed to a first-order traveling electron which encountered magnetic transition and the change in volume does not affect the crystal symmetry. The compositional effects caused to loss the travelling electron-metamagnetic transition and also lead to SOPT [29]. This even decreases the magnetocaloric effect and indicates the irrelevant loss of hysteresis.

The system of Gd_5(Si, Ge)$_4$ which shows the MCE and related to the RE5X4 family, encounters a magneto-structural first ordered phase evolution between orthorhombic Gd_5Si_4/Sm_5Ge_4 polymorphic structures. These structures are close to the Gd_5Si_4 and Gd_5Ge_4 binary compounds consisting of relative structures. Besides, the monoclinic-structures exist in Si content range of 0.2 - 0.4 in the pseudo binary $Gd_5(Si_xGe_{1-x})_4$ system. Therefore, the sequentially mentioned temperatures must be taken into an account: The monoclinic Curie temperature, orthorhombic Curie temperatures and the temperature (T_E) where both monoclinic and orthorhombic structures have equal free energies. The relationship between the T_C of orthorhombic$>T_E>T_C$ of monoclinic indicates the first ordered phase transition, which causes a pointed alteration in the magnetization. This is boosted as the increase in variance between orthorhombic T_C and T_E. The fractional replacement of Nb in $Gd_5Si_{2-x}Ge_{2-x}Nb_{2x}$ composition enlarged the T_c and magnetocaloric signal as 'x' rose to x=0.05 (monoclinic T_C around 295 K and ΔS $= -9.6$ $Jkg^{-1}K^{-1}$ for H=2T). Moreover, the smaller amount of Gd_5Si_3 phase is found in this case [23]. Most of the difficulties faced in optimizing the structures of the family are related to their impurity phases and homogeneity like hexagonal-phase of Gd_5Si_3. That is, the material is anti-ferromagnetic and unfavorable to magnetocaloric effect. Generally, the high temperature annealing is operated to achieve the homogeneity of the alloy. A strong improvement of ΔS and T_{ad} is described for samples which are annealed in comparison with the samples without annealing. The increment has been attributed to the rearrangement of germanium and silicon atoms in their respective positions of the monoclinic-structure. This also causes to eliminate the reserved orthorhombic phases of the polymorphs [24]. The arc melting technique offers an advanced cooling rate which also increases the magnetocaloric effect [25]. This has been achieved by using high-pure gadolinium [26]. On the other hand, the ferromagnetic lanthanum manganites are extensively studied for the MCE in refrigeration. The researchers Yu *et al.* [27], for magnetocaloric determinations, introduced ferromagnetic perovskite manganite ceramics originating from $LaMnO_3$ (lanthanum manganite). Moreover, these materials exhibit semi-conducting nature and anti-ferromagnetic nature at the temperature of 150 K. In the lanthanum manganates, the oxygen deficiency and La shortage, or the replacement of a non-trivalent ion in the place of rare earth trivalent ions pointed the presence of manganese ions with mixed valence level. Further, the double-exchange mechanism leads to ferromagnetic coupling between Mn^{4+} and Mn^{3+} ions.

These manganites show the critical magnetic phase diagram with respect to the composition. Nearly, all manganites which are metallic in nature exhibit the ferromagnetic nature. Then, the double-exchange mechanism is increased by the transfer of electrons while in anti-ferromagnetic manganites, the anti-parallel connection in adjacent manganese ions causes to null value for bounding integral and an insulator-system. Nevertheless, (Fig. **3.1**) [34] can get much more complexes with other types of manganites like glassy insulators, anti-ferromagnetic insulators, anti-ferromagnetic metals, ferromagnetic insulators, most complicated and mixed-phase states, and canted magnetic structures. The motivation for such an extensive phenomenology is the minor disparity in between probable ground states owing to contention amongst various interactions, as listed in reference [28].

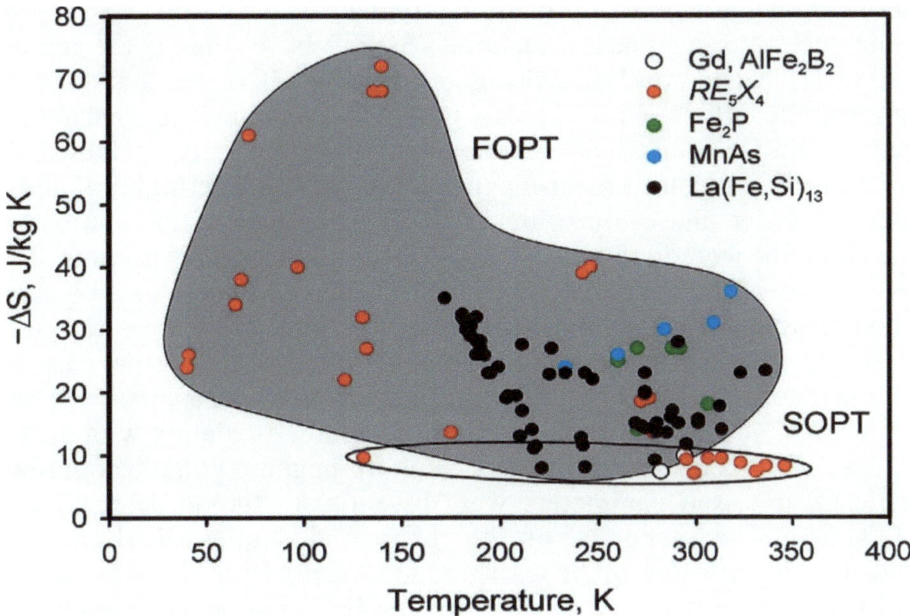

Fig. (3.1). Series of manganites exhibiting magnetic properties [34].

In addition, through generating an excess of oxygen or by substituting divalent ions (calcium, barium, strontium, *etc.*) or by substituting monovalent ions (sodium, potassium, lithium, *etc.*), one can enhance the Curie temperature over a broad range (from 150 to 375 K), in case of ferromagnetic manganites. Besides, the average radius of the cations at the dissimilar positions can also affect the ground-state of the manganites. These properties may be altered *via* partial replacement of lanthanum with yttrium, bismuth *etc.*, which are trivalent rare earth ions [27].

As above discussed alloys, the manganese-arsenic alloys have similar properties consisting of arsenic. The behavior of arsenic led studying the increase of the MCE properties of manganese-arsenic-alloys. Herein, the Mn-As alloys and rare earth elements are anticipated. On the one hand, the manganese family materials resulting from Mn-As and Fe_2P-type composites are also potential magnetocaloric materials. The Mn-As alloy displayed the first ordered transition at the temperature of 318 K from ferromagnetic to paramagnetic (hexagonal Ni-As to orthorhombic Mn-P) phase with massive change in the magnetic entropy (−30 $Jkg^{-1}K^{-1}$ for H = 5T) formerly with huge thermal hysteresis (6 K). This may be concentrated through fractional replacement of As for Sb (1 K). At 378 K, a

SOPT happens altering from Mn-P type to Ni-As type. Herein, the MnP-type structure is stable within the temperature range of 318 to 378 K [29]. Further, the manganese based alloys with fractional replacement of Sb [30, 31] lead to noticeable second order phase transition (SOPT) by changing the temperature from 220 to 318 K and preserving the huge values of MCE. The small transition is obtained for the alloys with increases of Mn content. The sharp transition observed for alloys with slight increment of manganese is attributed to non-ordinary travelling electron meta-magnetic transition while high amount of Mn surrenders an order phase transition of SOPT. At Curie temperature, this non-ordinary nature is accredited to a sturdy reduction in magnetostriction. Recently, the pressure dependence of magnetic characteristics of $MnAs_{1-x}Sb_x$ [32] detected a SOPT that pressures weakening the first-order behavior of the transition for x = 3. In the last five years, various research papers reported on the special case of MCE that is named colossal MCE for Mn-As at higher pressures. That bis it occurred for H=5 T, ΔS=−40 Jkg^{-1}K^{-1} at atmospheric pressure, escalating with pressure up to −267 Jkg^{-1}K^{-1} at 2.23 kbar. This was result of magneto volume coupling while unrealistic Gr uneissen parameters were essential to fit the detected points in MCE [33, 35]. The outcomes of the theoretical studies *via* Landau theory specified that the colossal MCE was just an object owing to estimate ΔS from nonequilibrium data deprived of the mixed-phase region or irreversibility of typical of FOPTs [34].

The family of Fe-Rh alloys in the form of unit ratio mixer was reported by the co researchers Manekar *et al.* [37], for the refrigeration applications. These were the first system for which GMCE was described [37]. Besides, the deficiency of reproduction of the original outcomes was taken as an irreversible nature of GMCE in this system [36]. To overcome this, Roy [37], demonstrated that the variation of magnetic field cycle at high temperatures should be isothermal for refrigeration cycle in this system. They also recommended the opportunity that the nearby variation of M-H curve leads to irreversibility of the MCE within the Fe-Rh alloys. The presence of MCE is a property of other systems showing a FOPT as explained above for the MnFe(P, Si, Ge) family Table **3.1**. The double Fe-Rh phase-diagram displays a ferromagnetic ordered CsCl structure for Rh component above 11% at 273 K.

Table 3.1. Ceramic materials exhibiting MCE.

Material	ΔS_{iso} (J kg—1K—1)	ΔT_{ad}(K)	Ref.
Dy5Si3.5Ge0.5	9.5		[37]
Dy5Si3Ge	34		[37]
MnFeP0.45As0.55	14.5/18		[38, 39]

(Table 3.1) cont.....

Material	ΔS_{iso} (J kg—1K—1)	ΔT_{ad}(K)	Ref.
MnFeP0.47As0.53	13		[39]
Mn1.1Fe0.9P0.47As0.53	20		[39]
MnFeP0.67Si0.22Ge0.11	14/27		[40]
MnFeP0.65Si0.24Ge0.11	16		[40]
MnFeP0.63Si0.26Ge0.11	27-Sep		[41]
MnFeP0.59Si0.30Ge0.11	13/27		[41]
MnFeP0.56Si0.33Ge0.11	25-Sep		[41]
MnFeP0.52Si0.48	10		[42]
MnFeP0.44Si0.56	9		[42]
MnAs-based	38/41, 36a	4.8/13	[43, 44]
MnAs0.90Sb0.10	24/30		[44]
MnAs0.85Sb0.15	26		[45]
MnAs0.25Sb0.25	18/24	5.5/10	[45]
Gd	5/9.8	5.7/11.5	[46]
AlFe2B2	4.4/7.3	1.8/3.0	[47]
RE5(Si,Ge)4-based			[48]
Gd5Si4	4.2/8.2	–/8.8	[48, 49]
Gd5Si3Ge	~8.7	8.6	[48]
Gd5Si2.5Ge1.5	~9.4	8.3	[48]
Gd5Si2Ge2		7.4/15.2	[46]
Gd5Si1.72Ge2.28	39	18.8	[48, 49]
Gd5SiGe3	68	11.8	[48, 49]
Gd5Si0.33Ge3.67	60.5	11.2	[48, 49]
Gd4.5Eu0.5Ge4	27	4.9	[50]

CONCLUSIONS

The magnetocaloric effect can become the main option for the refrigeration process in future. The isothermal entropy and adiabatic variation are the main properties to confine the refrigeration process. The magnetic materials such as Mn doped Fe alloys, rare earth elements such as Gd, La, Ce *etc.*, mixed composites and ceramics show the efficient thermal properties for the MCE which are presented successesfully in the chapter.

CONSENT FOR PUBLICATION

Not applicable.

CONFLICT OF INTEREST

The authors confirm that this chapter content has no conflict of interest.

ACKNOWLEDGEMENTS

The authors express thankfulness to Dr. P. Sreeramulu, Assistant Professor (English), GITAM, Bangalore for providing English language editing services to this manuscript.

REFERENCES

[1]　S. Qian, "Nasuta, D. Rhoads, A. Wang, Y. Geng, Y. Hwang, Y Radermacher, R. Takeuchi, I. Not-i--kind cooling technologies: A quantitative comparison of refrigerants and system performance", *Int. J. Refrig.*, vol. 62, pp. 177-192, 2016.
[http://dx.doi.org/10.1016/j.ijrefrig.2015.10.019]

[2]　P. Bansal, E. Vineyard, and O. Abdelaziz, "Status of not-in-kind refrigeration technologies for household space conditioning, water heating and food refrigeration", *Int. J. Sustain. Built Environ.*, vol. 1, pp. 85-101, 2012.
[http://dx.doi.org/10.1016/j.ijsbe.2012.07.003]

[3]　A. Kitanovski, U. Plaznik, U. Tomc, and A. Poredos, "Present and future caloric refrigeration and heat-pump technologies", *Int. J. Refrig.*, vol. 57, pp. 288-298, 2015.
[http://dx.doi.org/10.1016/j.ijrefrig.2015.06.008]

[4]　US department of energy of energy efficiency and renewable energy., "Using magnets to keep cool: Breakthrough technology boosts energy efficiency of refrigerators", Available online: https://www.energy.gov/eere/articles/using-magnets

[5]　US Department of energy—once of energy efficiency and renewable energy., "ORNL refrigerator cools with magnetism, not freon", Available online: https://www.energy.gov/ eere/buildings/ articles/orn

[6]　D. Vuarnoz, and T. Kawanami, "Numerical analysis of a reciprocating active-magnetic regenerator made of gadolinium wires", *Appl. Therm. Eng.*, vol. 37, p. 388, 2012.
[http://dx.doi.org/10.1016/j.applthermaleng.2011.11.053]

[7]　M.S. Kamran, J. Sun, Y.B. Tang, Y.G. Chen, J.H. Wu, and H.S. Wang, "Numerical investigation of room temperature magnetic refrigerator using micro channel regenerators", *Appl. Therm. Eng.*, vol. 102, pp. 1126-1140, 2016.
[http://dx.doi.org/10.1016/j.applthermaleng.2016.02.085]

[8]　T.F. Petersen, "Pryds, N. Smith, A. Hattel, J. Schmidt, H. Hogaard Knudsen, H.J. Two-dimensional mathematical model of a reciprocating room-temperature Active Magnetic Regenerator", *Int. J. Refrig.*, vol. 31, pp. 432-443, 2008.
[http://dx.doi.org/10.1016/j.ijrefrig.2007.07.009]

[9]　G.V. Brown, "Magnetic Stirling Cycles—A new application for magnetic materials", *IEEE Trans. Magn.*, vol. 13, pp. 1146-1148, 1977.
[http://dx.doi.org/10.1109/TMAG.1977.1059542]

[10]　V.K. Pecharsky, and K.A. Gschneidner, "Magnetocaloric effect from indirect measurements:

Magnetization and heat capacity", *J. Appl. Phys.,* vol. 86, pp. 565-575, 1999.
[http://dx.doi.org/10.1063/1.370767]

[11]   S.Y. Dan'kov, A. Tishin, V. Pecharsky, and K. Gschneidner, "Magnetic phase transitions and the magnetothermal properties of gadolinium", *Phys. Rev. B Condens. Matter Mater. Phys.,* vol. 57, pp. 3478-3490, 1998.
[http://dx.doi.org/10.1103/PhysRevB.57.3478]

[12]   J.S. Lee, "Evaluation of the magnetocaloric e_ect from magnetization and heat capacity data", *Phys. Status Solidi Basic Res,* vol. 241, pp. 1765-1768, 2004.
[http://dx.doi.org/10.1002/pssb.200304685]

[13]   K.K. Nielsen, H.N. Bez, L. von Moos, R. Bjørk, D. Eriksen, and C.R.H. Bahl, "Direct measurements of the magnetic entropy change", *Rev. Sci. Instrum.,* vol. 86, no. 10, 2015.103903
[http://dx.doi.org/10.1063/1.4932308] [PMID: 26520967]

[14]   V. Franco, "Blazquez, J.S. Ipus, J.J. Law, J.Y. Moreno-Ramírez, L.M. Conde, A. Magnetocaloric effect: From materials research to refrigeration devices", *Prog. Mater. Sci.,* vol. 93, pp. 112-232, 2018.
[http://dx.doi.org/10.1016/j.pmatsci.2017.10.005]

[15]   J. Tušek, A. Kitanovski, I. Prebil, and A. Poredoš, "Dynamic operation of an active magnetic regenerator (AMR): Numerical optimization of a packed-bed AMR", *Int. J. Refrig.,* vol. 34, pp. 1507-1517, 2011.
[http://dx.doi.org/10.1016/j.ijrefrig.2011.04.007]

[16]   M.Yu. Liu, "B. Numerical investigations on internal temperature distribution and refrigeration performance of reciprocating active magnetic regenerator of room temperature magnetic refrigeration", *Int. J. Refrig.,* vol. 34, pp. 617-627, 2011.
[http://dx.doi.org/10.1016/j.ijrefrig.2010.12.003]

[17]   C. Aprea, "Maiorino, A. A flexible numerical model to study an active magnetic refrigerator for near room temperature applications", *Appl. Energy,* vol. 87, pp. 2690-2698, 2010.
[http://dx.doi.org/10.1016/j.apenergy.2010.01.009]

[18]   K.A. Gschneidner Jr, and V.K. Pecharsky, "Magnetocaloric materials", *Annu. Rev. Mater. Sci.,* vol. 30, pp. 387-429, 2000.
[http://dx.doi.org/10.1146/annurev.matsci.30.1.387]

[19]   W. Tang, W. Lu, and X. Luo, "Particle size effects on La0.7Ca0.3MnO3: Size-induced changes of magnetic phase transition order and magnetocaloric study", *J. Magn. Magn. Mater.,* vol. 322, pp. 2360-2368, 2010.
[http://dx.doi.org/10.1016/j.jmmm.2010.02.038]

[20]   FC Medeiros, VD Mello, AL Dantas, FHS Sales, and AS Carrico, "Giant magnetoceloric effect of thin Ho films", *J. Appl. Phys,* vol. 109, p. 07A914, 2011.

[21]   B.G. Shen, J.R. Sun, F.X. Hu, H.W. Zhang, and Z.H. Cheng, "Recent progress in exploring magnetoceloric materials", *Adv. Mater.,* vol. 21, pp. 4545-4564, 2009.
[http://dx.doi.org/10.1002/adma.200901072]

[22]   F.X. Hu, B.G. Shen, J.R. Sun, Z.H. Cheng, and G.H. Rao, "Zhang. Influence of negative lattice expansion and metamagnetic transition on magnetic entropy change in the compound LaFe11.4Si1.6", *Appl. Phys. Lett.,* vol. 78, pp. 3675-3677, 2001.
[http://dx.doi.org/10.1063/1.1375836]

[23]   M. Saoussen, B. Mohamed, M. Rafik, E.K. Hlil, and O. Mohamed, "Magnetocaloric study and estimation of the spontaneous magnetization by a magnetic entropy analysis in Pr0.6Ca0.1Sr0.3Mn0.975Fe0.025O3", *J. Alloys Compd.,* vol. 680, pp. 381-387, 2016.
[http://dx.doi.org/10.1016/j.jallcom.2016.04.113]

[24]   A.O. Pecharsky, K.A. Gschneidner Jr, and V.K. Pecharsky, "The giant magnetocaloric effect of optimally prepared Gd5Si2Ge2", *J. Appl. Phys.,* vol. 93, pp. 4722-4728, 2003.

[http://dx.doi.org/10.1063/1.1558210]

[25]   A. Yan, A. Handstein, P. Kerschl, K. Nenkov, K.H. Muller, and O. Gutfleisch, "Effect of composition and cooling rate on the structure and magnetic entropy change in Gd5SixGe4−x", *J. Appl. Phys.,* vol. 95, pp. 7064-7066, 2004.
[http://dx.doi.org/10.1063/1.1667852]

[26]   K.A. Gschneidner Jr, and V.K. Pecharsky, "Magnetic refrigeration materials (invited)", *J. Appl. Phys.,* vol. 85, pp. 5365-5368, 1999.
[http://dx.doi.org/10.1063/1.369979]

[27]   M. Phan, and S. Yu, "Review of the magnetocaloric effect in manganite materials", *J. Magn. Magn. Mater.,* vol. 308, pp. 325-340, 2007.
[http://dx.doi.org/10.1016/j.jmmm.2006.07.025]

[28]   K. Dorr, "Ferromagnetic manganites: Spin-polarized conduction *versus* competing interactions", *J. Phys. D Appl. Phys.,* vol. 39, pp. R125-R150, 2006.
[http://dx.doi.org/10.1088/0022-3727/39/7/R01]

[29]   H. Wada, and Y. Tanabe, "Giant magnetocaloric effect of MnAs1−xSbx", *Appl. Phys. Lett.,* vol. 79, pp. 3302-3304, 2001.
[http://dx.doi.org/10.1063/1.1419048]

[30]   Z.G. Xie, D.Y. Geng, and Z.D. Zhang, "Reversible room-temperature magnetocaloric effect in Mn5Pb2", *Appl. Phys. Lett.,* vol. 97, 2010.202504
[http://dx.doi.org/10.1063/1.3518064]

[31]   T. Morikawa, "WadaH, KogureR,Hirosawa S. Effect of concentration deviation from stoichiometry on the magnetism of MnAsSb", *J. Magn. Magn. Mater.,* vol. 283, pp. 322-328, 2004.
[http://dx.doi.org/10.1016/j.jmmm.2004.05.035]

[32]   H. Wada, S. Matsuo, and A. Mitsuda, "Pressure dependence of magnetic entropy change and magnetic transition in MnAs1−xSbx", *Phys. Rev. B Condens. Matter Mater. Phys.,* vol. 79, 2009.092407
[http://dx.doi.org/10.1103/PhysRevB.79.092407]

[33]   A. de Campos, D.L. Rocco, A.M. Carvalho, L. Caron, A.A. Coelho, S. Gama, L.M. da Silva, F.C. Gandra, A.O. dos Santos, L.P. Cardoso, P.J. von Ranke, and N.A. de Oliveira, "Ambient pressure colossal magnetocaloric effect tuned by composition in Mn(1-x)Fe(x)As", *Nat. Mater.,* vol. 5, no. 10, pp. 802-804, 2006.
[http://dx.doi.org/10.1038/nmat1732] [PMID: 16951675]

[34]   J.S. Amaral, "Amaral VS. The effect of magnetic irreversibility on estimating the magnetocaloric effect from magnetization measurements", *Appl. Phys. Lett.,* vol. 94, 2009.042506
[http://dx.doi.org/10.1063/1.3075851]

[35]   DT Cam Thanh, E Bruck, O Tegus, JCP Klaasse, TJ Gortenmulder, and KHJ Buschow, "Magnetocaloric effect in Fe(P,Si,Ge) compounds", *J. Appl. Phys,* vol. 99, p. 08Q107, 2006.

[36]   T.J. Zhou, Z. Yu, W. Zhong, X.N. Xu, H.H. Zhang, and Y.W. Du, "Larger magnetocaloric effect in two-layered La1.6Ca1.4Mn2O7La1.6Ca1.4Mn2O7 polycrystal", *J. Appl. Phys.,* vol. 85, pp. 7975-7977, 1999.
[http://dx.doi.org/10.1063/1.369387]

[37]   M. Manekar, and S.B. Roy, "Reproducible room temperature giant magnetocaloric effect in Fe-Rh", *J. Phys. D Appl. Phys.,* vol. 41, 2008.192004
[http://dx.doi.org/10.1088/0022-3727/41/19/192004]

[38]   E. Brueck, and M. Ilyn, "Tishin, A. M. Tegus, O. Magnetocaloric effects in MnFeP1−xAsx-based compounds", *J. Magn. Magn. Mater.,* vol. 290–291, no. Pt. 1, pp. 8-13, 2005.
[http://dx.doi.org/10.1016/j.jmmm.2004.11.152]

[39]   Thanh Cam, Bruck D.T, Tegus E., Klaasse O., Gortenmulder J.C.P, and Buschow T.J, K. H. J, "Magnetocaloric effect in MnFe(P, Si, Ge) compounds", *J. Appl. Phys,* vol. 99, no. 8, p. 08Q107,

2006.

[40] D.T.C. Thanh, Bruck, E. Trung, N. T. Klaasse, J. C. P. Buschow, K. H. J.; Ou, Z. Q.; Tegus, O. Caron, L., *J. Appl. Phys.,* vol. 103, no. 7, Pt. 2, 2008.

[41] H. Wada, and Y. Tanabe, "Appl. The isothermal variation of the entropy (ΔST) may be miscalculated from magnetization isotherms in some cases: MnAs and Gd5Ge2Si2 compounds as examples", *Phys. Lett.,* vol. 79, no. 20, pp. 3302-3304, 2001.

[42] D. T. C. Thanh, E. Bruck, and N. T. Trung, Klaasse, J. C. P. Buschow, K. H. J.; Ou, Z. Q.; Tegus, O. Caron, L., *Appl. Phys. Magnetocaloric Effect and Magnetocaloric Materials,* vol. 103, no. 7, Pt. 2, 2008.

[43] A.M.G. Carvalho, A.A. Coelho, P.J. von Ranke, and C.S. Alves, "The isothermal variation of the entropy (ΔST) may be miscalculated from magnetization isotherms in some cases: MnAs and Gd5Ge2Si2 compounds as examples", *J. Alloys Compd.,* vol. 509, no. 8, pp. 3452-3456, 2011.
[http://dx.doi.org/10.1016/j.jallcom.2010.12.088]

[44] H. Wada, T. Morikawa, K. Taniguchi, T. Shibata, Y. Yamada, and Y. Akishige, "Giant magnetocaloric effect of MnAs1−xSbx in the vicinity of first-order magnetic transition", *Physica B,* vol. 328, no. 1–2, pp. 114-116, 2003.
[http://dx.doi.org/10.1016/S0921-4526(02)01822-7]

[45] H. Wada, Y. Tanabe, K. Hagiwara, and M. Shiga, "Magnetic phase transition and magnetocaloric effect of DyMn2Ge2", *J. Magn. Magn. Mater.,* vol. 218, pp. 203-210, 2000.
[http://dx.doi.org/10.1016/S0304-8853(00)00410-8]

[46] V.K. Pecharsky, "Gschneidner, K. A., Giant Magnetocaloric Effect in Gd5(Si2Ge2) Jr", *Phys. Rev. Lett.,* vol. 78, no. 23, pp. 4494-4497, 1997.
[http://dx.doi.org/10.1103/PhysRevLett.78.4494]

[47] X. Tan, P. Chai, C.M. Thompson, and M. Shatruk, "Magnetocaloric effect in AlFe2B2: toward magnetic refrigerants from earth-abundant elements", *J. Am. Chem. Soc.,* vol. 135, no. 25, pp. 9553-9557, 2013.
[http://dx.doi.org/10.1021/ja404107p] [PMID: 23731263]

[48] K.A. Gschneidner Jr, and V.K. Pecharsky, "Annu.magnetocaloric materials", *Rev. Mater. Sci,* vol. 30, pp. 387-429, 2000.
[http://dx.doi.org/10.1146/annurev.matsci.30.1.387]

[49] Y.I. Spichkin, V.K. Pecharsky, and K.A. Gschneidner, "Preparation, crystal structure, magnetic and magneto thermal properties of (GdxR5−x) Si4, (GdxR5−x) Si4, where R=Pr and Tb, alloys", *J. Appl. Phys.,* vol. 89, no. 3, pp. 1738-1745, 2001.
[http://dx.doi.org/10.1063/1.1335821]

[50] J. Yao, "Wang, P. L. Mozharivsky, j,Y.J.Crystal structure and magnetism of Gd5−xEuxGe4", *J. Alloys Compd.,* vol. 534, pp. 74-80, 2012.
[http://dx.doi.org/10.1016/j.jallcom.2012.04.051]

CHAPTER 4

Advanced Ceramics for Thermoelectric Power Generation

S. Ramesh[1,*], **K. Chandra Babu Naidu**[1,*], **N.V. Krishna Prasad**[1], **N. Suresh Kumar**[2], **A. Mallikarjuna**[1], **K. Venkata Ratnam**[3] and **H. Manjunatha**[3]

[1] *Department of Physics, GITAM Deemed to be University, Bengaluru-562163, Karnataka, India*

[2] *Department of Physics, JNTU, Anantapuramu-515002, A.P, India*

[3] *Dept.of Chemistry, GITAM Deemed to be University, Bangalore-562163, India*

Abstract: In the future, a lot of problems can be faced in energy management due to the heavy consumption of renewable and non-renewable energy sources. This is because of the ever-growing population and huge usage of electronic appliances in modern life. Therefore, people are looking into different fields to increase energy demand. In this chapter, the role of advanced ceramics and their applications of thermoelectric technology were discussed. Also, an attempt was made to study different advanced ceramic materials that are used for thermoelectric power generation in recent years.

Keywords: Applications, Thermal Conductivity, Thermal Efficiency, Seebeck effect, ZT.

4.1. INTRODUCTION

Advanced ceramics are specified as the new class of ceramic materials made up of high purity synthetic chemicals. These materials are focused on different industrial applications due to their high performance. In addition, they are composed of oxides, carbides and nitrides. Accordingly, the advanced ceramic materials based on their compositions are classified into two categories oxide or non-oxide ceramics.

• Oxide ceramics include binary oxides, aluminates, ferrites, titanites, niobates, zirconates, *etc.*

• Non-oxide ceramics cover carbides, nitrides, borides, carbon, *etc.*

* **Corresponding author Dr. K. Chandra Babu Naidu:** GITAM Deemed To Be University-Bangalore Campus, Bangalore-562163, Karnataka, India; E-mails: chandrababu954@gmail.com & sramesh664@gmail.com

• Ceramic composites include both the combinations of oxides and non-oxides.

Ceramic materials are a different class of materials composed of non-metallic stable and inorganic materials. They are brittle, electrically and thermally insulated materials prepared with a combination of more than a single element. The mechanical strength of these materials can be increased by sintering process [1]. The ceramic materials will have distinct properties due to their grain boundaries in the materials. These grain boundaries undergo misalignments with nearby grains and their variations in the structure. These variations can be the quality in the perfections, contributions along with shape, size and internal stress to which they are exposed. For the last few decades, scientists have observed the tremendous increase in properties like ferroelectric, ferromagnetic, pyroelectric, dielectric and magnetoresistance, superconducting and gas sensing applications. The chemical and thermal properties of oxide and non-oxide ceramics were increased by the different manufacturing techniques. These ceramic materials also became the main materials for advanced technologies like energy transformation, storage, supply, in the manufacturing, medical field technology, transportation and information technology as shown in Fig. (**4.1**).

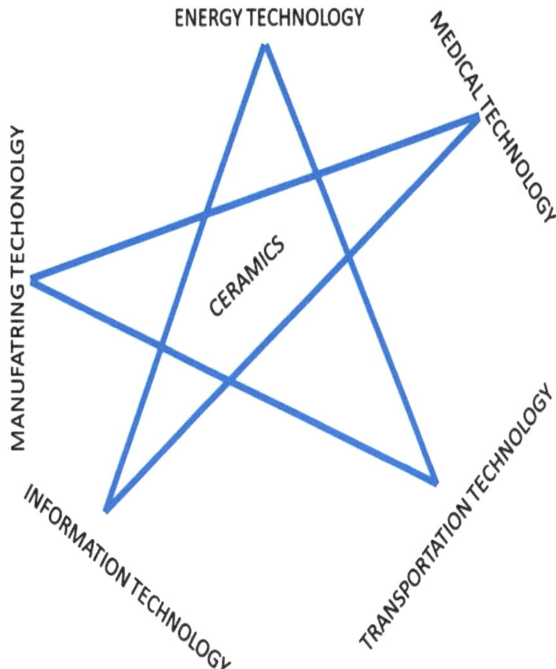

Fig. (4.1). Applications of advanced ceramics in various fields of technology.

4.2. ADVANCED CERAMICS AND THEIR PROPERTIES

4.2.1. Brittleness

The ceramic materials can sustain the brittleness up to a certain high temperature. This is happened because of the ceramic compounds containing the mixed ionic covalent bonds that dominant the division of atoms together. But they are brittle in nature at the room temperature.

4.2.2. Strength

Ceramic compounds, in general, will have very poor toughness; despite of increasing the toughness by adding the composites to a particular ceramic material. Ice-templating is a technique used to increase the mechanical strength of the ceramic material by subjecting it to the considerable mechanical loading. This maintains the microstructure of the ceramics and especially, the mechanical strength of the ceramic material. This property is very important for the applications like solid oxide fuel cells and water filtration.

4.2.3. Transparency

Due to large energy gap (E_g) values, few of the ceramic materials are transparent like glass optical fiber having the transparency of 96%. The ceramics which exhibit the transparency in nature cover the materials made up of sapphire, few valuable stones and optical fiber.

4.2.4. Chemical Insensitivity

The ceramic Pyrex glass is extensively used for preparation of bakeware and in chemical labs due to the high resistant nature of many destructive chemicals. In addition, these chemicals are of durable in nature at very large temperatures and are high resistant to the thermal shocks due to low coefficient of thermal expansion of value of about 33×10^{-7} K^{-1}.

4.2.5. Electrical and Thermal Influence

We know that diamond is a carbon material having highest thermal conductivity of the order of 1700 W/mK. Herein, the mechanism of the conduction is developed due to the phonons but not due to the electrons. In case of conductors, the valence electrons are free to move through the lattices, whereas in case of ceramics, they are tightly connected to the bonds. Hence, they will have very low thermal and electrical conduction. These ceramics can also have high thermal conductivity. In particular, (i) ReO_3 is the ceramic material which has the electrical conductivity similar to the copper (Cu) and (ii) the ceramic material in the form of mixed oxide ($YBa_2Cu_3O_7$) exhibits the superconductivity and zero resistivity below 92 K.

4.3. THERMOELECTRIC CERAMIC MATERIALS AND REQUIRED PARAMETERS

Among sevral materials, only few are considered as thermoelectric materials. Thermoelectric materials are classified based on the temperature range of operation.

• Bismuth based alloys on mixing with the antimony, tellurium or selenium are referred as low temperature thermoelectric materials with a temperature range of 450 K.

• Lead based alloys obtained an intermediate temperature range of about 850 K.

• The high temperature thermoelements are fabricated with Silicon-Germanium alloys over a temperature range of 1300 K.

Now, there is great trend in the research field of thermoelectric power generation in order to improve the power generation and the material efficiency. The efficiency of thermoelectric materials can be calculated by dimensionless figure of merit, ZT.

$$ZT = \frac{S^2 \sigma T}{k} \tag{4.1}$$

In Eq. (1) σ refers to electrical conducting, T refers to temperature and k refers to thermal conductivity while S represents the Seebeck coefficient. Yet, the amount of heat transfer by electrons and the phonons influence the thermal conductance. These ceramic materials show relatively high (Figure of merit) value. In this context, these materials have a tendency of limited chemical stability, toxicity and high cost. Hence, these materials are not suitable for the construction of thermoelectric power generation. The figure of merit *versus* conversion efficiency plot at operating temperature difference is dipicted in the Fig. **(4.2)** [2]. Obviously, as prescribed by the Carnot efficiency, the increase in temperature difference brings a comparable gain in the feasible heat conversion. Therefore, the high temperature differences are appreciable.

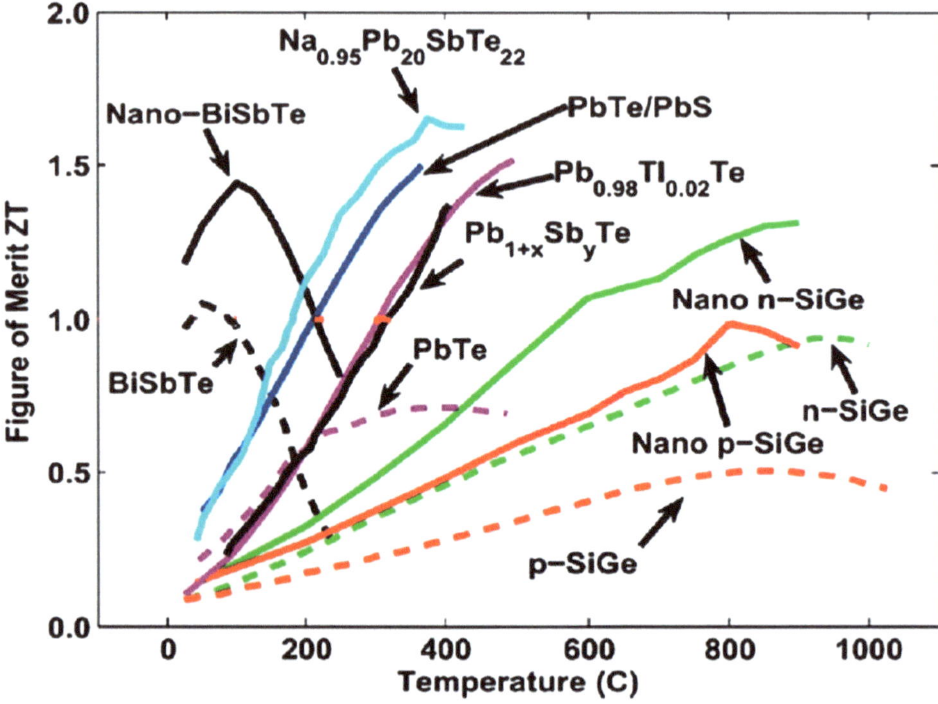

Fig. (4.2). ZT for different thermoelectric materials and its applications [2].

Wiedemann-Franz law [3] describes the above relation as:

$$LT = \frac{Ke}{\sigma} \tag{4.2}$$

For the efficient thermoelectric material to come into the field of power generation, efficiency of thermoelectric material is to be increased at high temperatures. It is shown that the thermoelectric devices with value 2 (figure of merit) are feasible for several applications. If the figure of merit is 3, they can grant the thermoelectric generators to challenge along with the conventional mechanical power generators [4]. The passage of electric current in the ceramic materials is given by the electrical conductivity equation:

$$\sigma = \frac{1}{\rho} = \frac{1}{RA} \qquad (4.3)$$

Power generation performance calculation of the ceramic material is:

$$Z = \frac{\alpha^2}{kR} \quad \text{and } \alpha = -\frac{\Delta V}{\Delta T}$$

$$ZT = \frac{\alpha^2 T}{kR} \quad \text{and } T = \frac{T_H + T_L}{2} \qquad (4.4)$$

Fig. (**4.3**) [5] indicates the usage of thermoelectric materials either for power generation or cooling purpose. The construction of heat engines comprises of collection of semiconductors (N & P) having heat source and sink on either side by producing electric power. Electric power can be altered for cooling or heating by changing the current direction.

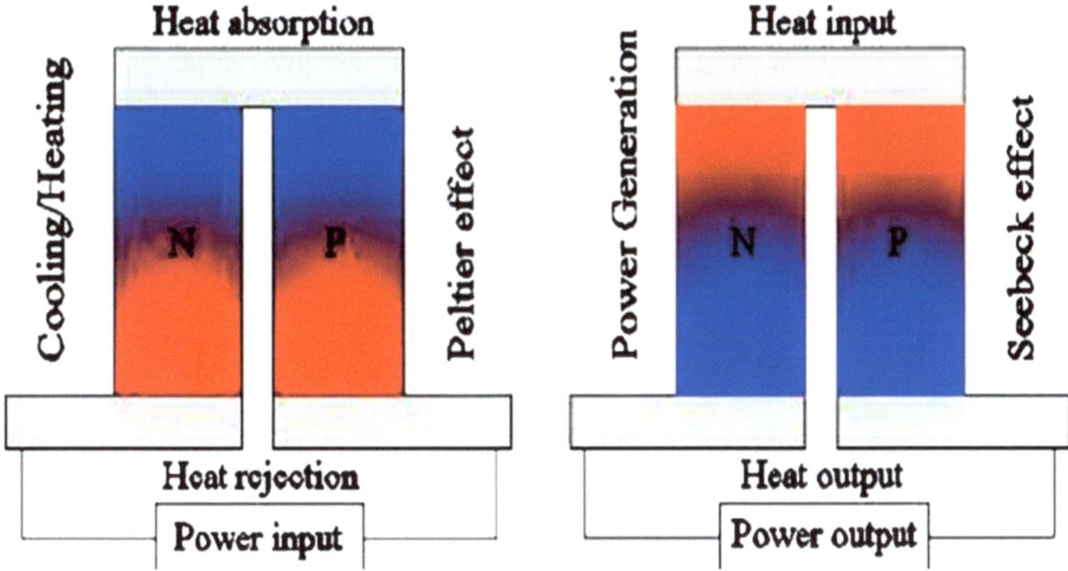

Fig. (4.3). Usage of thermoelectric materials for either cooling or power generation [5].

The maximum conversion efficiency is of the Carnot's engine is:

$$\eta = \eta_{carnot} \left[\frac{\sqrt{1+ZT}-1}{\sqrt{1+ZT}+\dfrac{T_L}{T_H}} \right] \qquad (4.5)$$

4.4. THERMOELECTRIC CERAMIC MATERIALS

There are some common thermoelectric ceramic materials such as bismuth telluride (Bi_2Te_3), plumbous telluride (PbTe), silicon germanium (SiGe) and thier alloys. The bismuth telluride (Bi_2Te_3) has almost larger molecular weight, density of 7.86 g/cm^3 and melting point of 585°C. If the concentration of Bi is more in bismuth telluride, it can be used as a P-type material. Similarly, if it is having more Te, bismuth telluride is treated as n-type material [6]. The Seebeck coefficients of p-type and n-type materials are 185 µV/K and -205 µV/K and thermal conductivity (k) is about 1.9 W/(m.K), respectively [7].The PbTe alloys are having cubic structure having melting point about 922°C and bang gap of 0.30 eV which is double that of (Bi_2Te_3). It is mostly used for the power generation in the range of temperature about 300-900 K [8]. The thermal conductivity is lowered mostly for the materials of solid solution formed with Si and Ge due to its high temperature thermoelectric behavior. Currently, the used solid solutions are $Si_{0.70}Ge_{0.30}$, $Si_{0.80}Ge_{0.20}$ and $Si_{0.8}5Ge_{0.15}$ which showed optimum values of SiGe alloy *versus* temperature [9].

Table 4.1. Different thermoelectric materials and their thermal efficiency.

S.No	Ceramic Materials	T_{hot} (K)	(K)	Power Input (Output)	Thermal Efficiency	Purpose	Ref.
1.	Bi_2Te_3, PbTe	373	1073	0.5–1 kW	5-10%	Electricity generation	[10, 11]
2.	PbTe, SiGe	366	783 -1300	25-56W	---	Electricity generation	[12]
3.	SiGe	293	871	46.8W	4.4%	Electricity generation	[13, 14]
4.	Bi_2Te_3, PbTe	303	473-911	4W-1kW	4%	Electricity and hot water generation	[15 - 18]
5.	ZnSb, Bi_2Te_3	----	85	19.13pW	---	---	[19]

(Table 4.1) cont.....

S.No	Ceramic Materials	T_{hot} (K)	(K)	Power Input (Output)	Thermal Efficiency	Purpose	Ref.
6.	CdZnTe	283	313	3W	---	---	[20 - 24]
7.	$CoSb_3$	850	300		14.7	---	[25]
8.	Bi_2Te_3, SiGe	----	35-1000	0.04-17000 [μW/mm^2]	0.1-2	Thermoelectric generator	[26]
9.	SiC	721	421	1.708 W/m^2	---	Aerodynamic heat recovery	[27]

4.5. BI_2TE_3 BASED CERAMICS

The study of thermoelectric properties of Bi_2Te_3 was started long back in 1954. At room temperature, Bi_2Te_3 is a kind of classical material which acts as sophisticated material for wide range of applications. The indirect band gap of the same material is about 0.15 eV. Strong covalent bond exists between Bi and Te atoms while the Vander Waal bonding is weak. In addition, the Bi_2Te_3 is very convenient material to make P and N-type with Sb and Se materials which are preferable for manufacturing the thermoelectric devices. Using the melt spinning-spark plasma sintering method, scientists fabricated $Bi_{0.48}Sb_{1.52}Te_3$ with multiscale microstructure (Nano-crystals having coherent grain boundaries) which are sufficiently having low thermal conductivity. A very high value of ZT of 1.5 was gained for the above samples. Scientists reported a challenging value for ZT (1.86) in $Bi_{0.5}Sb_{1.5}Te_3$ which was developed with the help of melt spinning-spark plasma sintering. In different volumes, graphene is homogeneously spread in nano Bi_2Te_3 through different combinations like magnetic stirring, ultra-sonication and ball milling. The electrical conductivity of the composites is increased by adding of graphene. The enhancement in power factor and the decrease in thermal conductivity generate exceptional increase in the figure of merit at 1.5% of graphene at 500 K [28]. The thermoelectric figure of merit ZT of $Bi_{0.4}Sb_{1.6}Te_3$ bulk materials mixed with various NH_4HCO_3 content showed the maximum figure of merit of 1.11 at temperature 343 K (Table **4.1**). Moreover, sample with the pre-mixing of NH_4HCO_3 shows the higher value of figure of merit compared to samples prepared without mixing of NH_4HCO_3 due to lower thermal conductivity and See-beck coefficient [29]. The n-type γ-Al_2O_3 – dispersed $Bi_2Se_{0.3}Te_{2.7}$ nano-compositions are studied for the similar properties. The alternate content of $Bi_2Se_{0.3}Te_{2.7}$-1.0 Vol %, γ-Al_2O_3 exhibits the high ZT of 0.99 at 400 K.

4.6. PBTE BASED CERAMICS

The PbTe sample shows similar structure as Sodium Chloride (NaCl) with a small band gap given by 0.33 eV. It can be easily doped into n and p-type materials. These are the excellent materials which can be adopted at moderate temperatures and in space application. Wide varieties of methods are used for increasing the performance of the PbTe. It showed the result of ZT with 1.5 for PbTe doped with Ti and found the increase of Seebeck coefficient due to deviation in the electron density. In addition, it was concluded that the convergence of electronic bands in $PbTe_{1-x}Se_x$ leads to high valley of degeneracy by continuing the peak value of ZT ~ 1.8 to a temperature of 850 K. Again in 2014, they conducted optimization for ZT of 1.2 in N-type material of PbTe doped with La and I. Scientists performed number of individual analysis and observed a peak value for ZT ~ 1.15 when doped with Al. This is determined to be the best relation for carrier concentration and thermoelectric performance [30]. In recent studies, it was observed that for reducing the thermal conductivity, they modified the grain, the dislocation density and got average grain strain of 1.3% in $Na_{0.03}Eu_{0.03}Sn_{0.02}Pb_{0.02}Te$ sample. Therefore, the lattice strain affects the interaction force between the atoms which reduces the thermal conductivity of the lattice. The high value of ZT of 2.6 was gained for $Na_{0.03}Eu_{0.03}Sn_{0.02}Pb_{0.02}Te$ which showed the high stability. A new rare division of BiTe-PbTe hybrid semiconductors of n-type and p-type form a hybrid module of superior performance and temperature stability. They are allowed to with stand up to a temperature of 360°C. The TEG1-PB class of materials are able to operate at a high temperature. These material surfaces are covered with graphite. They also work at 200°C to 360°C temperature range and offer a very high-power output over 260°C hot side.

4.7. SIGE BASED CERAMICS

SiGe based ceramics are other types of thermoelectric materials used at very high temperature of 1275 K and are used for the space exploration. Scientists indicated a ZT value of 1.3 in $Si_{95}Ge_5$ (with nano of $Si_{70}Ge_{30}P_3$ doped with nanocomposites particles). The B-doped nano-structured bulk $Si_{80}Ge_{20}$ fixed with SiO_2 nano insertion was developed by mechanical alloying and spark plasma sintering. It showed the decrease in thermal conductivity and increase in Seebeck coefficient with figure of merit 0.72. Research on the silicon, germanium materials (thermoelectric) are reduced in recent years due to expensive price of Ge and low dimensional materials like nanowires, thin films and superlattices.

CONCLUSIONS

The efficiency of the thermoelectric power generation depends on the figure of merit of the materials, low thermal conductivity and high electrical conductivity. We discussed various ceramic materials and their applications in the field of power generation. A wide research is going on thermoelectric materials by manipulating nanostructures of ceramic material to increases the Seebeck coefficient and decreasing the thermal conductivity.

CONSENT FOR PUBLICATION

Not applicable.

CONFLICT OF INTEREST

The authors confirm that this chapter content has no conflict of interest.

ACKNOWLEDGEMENTS

The authors express thankfulness to Dr. P. Sreeramulu, Assistant Professor (English), GITAM, Bangalore for providing English language editing services to this manuscript.

REFERENCES

[1]   W.D. Kingery, H.K. Bowen, and D.R Uhlmann, *Introduction to ceramics* Second edition. Interscience publication: Wiley. ISBN 0-471.47860-1

[2]   D. Zabek, and F. Morini, "Solid state generators and energy harvesters for waste heat recovery and thermal energy harvesting", *Therm. Sci. Eng. Prog,* vol. 9, pp. 235-247, 2019.
[http://dx.doi.org/10.1016/j.tsep.2018.11.011]

[3]   D.M. Rowe, Ed., *CRC Handbook of Thermoelectric.* CRC Press, 1995.
[http://dx.doi.org/10.1201/9781420049718]

[4]   C.J. Vineis, A. Shakouri, A. Majumdar, and M.G. Kanatzidis, "Nanostructured thermoelectrics: big efficiency gains from small features", *Adv. Mater.,* vol. 22, no. 36, pp. 3970-3980, 2010.
[http://dx.doi.org/10.1002/adma.201000839] [PMID: 20661949]

[5]   X.F. Zheng, C.X. Liu, Y.Y. Yann, and Q. Wang, "A review of thermoelectrics research – Recent developments and potentials for sustainable and renewable energy applications", *Renew. Sustain. Energy Rev.,* vol. 32, pp. 486-503, 2014.
[http://dx.doi.org/10.1016/j.rser.2013.12.053]

[6]   F.J. DiSalvo, "Thermoelectric cooling and power generation", *Science,* vol. 285, no. 5428, pp. 703-706, 1999.
[http://dx.doi.org/10.1126/science.285.5428.703] [PMID: 10426986]

[7]   T.M. Tritt, "Holey and unholy semiconductors", *Science,* vol. 283, pp. 804-805, 1999.
[http://dx.doi.org/10.1126/science.283.5403.804]

[8]   G.S. Nolas, D.T. Morelli, and T.M. Tritt, "A Phonon-glass-electron crystal approach to advanced thermoelectric energy conversion applications", *Annu. Rev. Mater. Sci.,* vol. 29, pp. 89-116, 1999.
[http://dx.doi.org/10.1146/annurev.matsci.29.1.89]

[9]   S.U. Tai-Chao, Z.H.U. Pin-Wen, and M.A. Hong, *Chinese Journal of High-Pressure Physics,* vol. 21, no. 1, pp. 55-58, 2007.

[10]   K. Matsubara, *Proceedings of the 15th international conference thermoelectrics,* 2002p. 418 CA

[11]   K. Matsubara, and M. Matsuura, *Thermoelectric applications to vehicles. Thermoelectric handbook, macro to nano.* CRC Press: Boca Raton, FL, 2005.

[12]   D.M. Rowe, *CRC handbook of thermoelectrics.* CRC Press: Boca Raton, FL, 1995.
[http://dx.doi.org/10.1201/9781420049718]

[13]   T. Kajikawa, T. Nishikaigan, and F. Kanagawa, "Thermoelectric power generation systems recovering heat from combustible solid waste in Japan", *Proceedings of the 15th international conference on thermoelectrics,* 1996
[http://dx.doi.org/10.1109/ICT.1996.553504]

[14]   "Kajikawa Thermoelectric power generation systems recovering heat from combustible solid waste in Japan", *Proceedings of the 15th international conference on thermoelectrics,* Pasadena, USA, pp. 343-51, 1996.

[15]   A. Killander, "A stove-top generator for cold areas", *Proceedings of the 15th international conference on thermoelectric,* 1996

[16]   G. Min, and D.M. Rowe, "Symbiotic application of thermoelectric conversion for fluid preheating/power generation", *Energy Convers. Manage.,* vol. 43, pp. 221-228, 2002.
[http://dx.doi.org/10.1016/S0196-8904(01)00024-3]

[17]   X.F. Zheng, Y.Y. Yan, and K. Simpson, "A potential candidate for the sustainable and reliable domestic energy generation-thermoelectric cogeneration system", *Appl. Therm. Eng.,* vol. 53, no. 2, pp. 305-311, 2013.
[http://dx.doi.org/10.1016/j.applthermaleng.2012.03.020]

[18]   K. Qiu, and A.C.S. Hayden, "A natural-gas-fired thermoelectric power generation system", *J. Electron. Mater.,* vol. 38, pp. 1315-1319, 2012.
[http://dx.doi.org/10.1007/s11664-008-0648-4]

[19]   P. Fan, Z. Zheng, Z. Cai, T. Chen, P. Liu, and X. Cai, "The high performance of a thin film thermoelectric generator with heat flow running parallel to film surface", *Appl. Phys. Lett.,* vol. 102, 2013.033904
[http://dx.doi.org/10.1063/1.4788817]

[20]   R.H. Redus, "Improved thermoelectrically cooled X/γ-ray detectors and electrobics. NuclInstrum Methods Phys Res Sect A: Accel, Spectrom", *Detect Assoc Equip,* vol. 458, pp. 214-219, 2001.
[http://dx.doi.org/10.1016/S0168-9002(00)00864-0]

[21]   G. Bale, "Cooled CdZnTe detectors for X-ray astronomy. NuclInstrum Methods, Phys Res Sect A: Accel, Spectrom", *Detect Assoc Equip,* vol. 436, pp. 150-154, 1999.
[http://dx.doi.org/10.1016/S0168-9002(99)00612-9]

[22]   M.K. Scruggs, *"Thermal-electrically cooled photodetector",* United States patent application, 20010032922, kind code, A1, 2001.

[23]   G. Min, "Cooling performance of integrated thermoelectric micro cooler", *Solid-State Electron.,* vol. 43, pp. 923-924, 1999.
[http://dx.doi.org/10.1016/S0038-1101(99)00045-3]

[24]   G.J. Snyder, and A.H. Snyder, "Figure of merit ZT of a thermoelectric device defined from materials properties", *Energy Environ. Sci.,* vol. 10, pp. 2280-2283, 2017.
[http://dx.doi.org/10.1039/C7EE02007D]

[25]   J. Hussam, K. Navid, A. Sulaiman, D. Bertrand, C. Amisha, and A.T. Savvas, "Waste heat recovery technologies and applications", *Therm. Sci. Eng. Prog.,* vol. 6, pp. 268-289, 2018.
[http://dx.doi.org/10.1016/j.tsep.2018.04.017]

[26]   X-Y. Han, J. Wang, and H-F. Cheng, "Investigation of thermoelectric SiC ceramics for energy harvesting applications on supersonic vehicles leading–edges", *Bull. Mater. Sci.,* vol. 37, pp. 127-132, 2014.

[http://dx.doi.org/10.1007/s12034-014-0613-1]

[27]   K. Ahmad, C. Wan, M.A. Al-Eshaikh, and A.N. Kadachi, "Enhanced thermoelectric performance of Bi_2Te_3 based graphene nanocomposites", *Appl. Surf. Sci.,* vol. 474, pp. 2-8, 2019.
[http://dx.doi.org/10.1016/j.apsusc.2018.10.163]

[28]   X. Hu, J. Hu, X. an Fan, B. Feng, Z. Pan, P. Liu, and Y. Li, "Artificial porous structure: An effective method to improve thermoelectric performance of Bi_2Te_3 based alloys", *J. Solid State Chem.,* vol. 282, 2019.121060
[http://dx.doi.org/10.1016/j.jssc.2019.121060]

[29]   Bowen Cai, Haihua Hu, Hua-Lu Zhuang, and Jing-Feng Li, "Promising materials for thermoelectric applications", *Journal of alloys and compounds,* vol. 806, pp. 471-486, 2019.

[30]   A. Usenko, D. Moskovshikh, M. Gorshenkov, A. Voronin, A. Stepashkin, S. Kaloshkin, D. Arkhinpov, and V. Khovaylo, "Enhanced thermoelectric figure of merit of p-type $Si_{0.8}Ge_{0.2}$ nanostructured spark plasma sintered alloys with embedded SiO_2 nanoinclusions", *Scr. Mater.,* vol. 127, pp. 63-67, 2017.
[http://dx.doi.org/10.1016/j.scriptamat.2016.09.010]

CHAPTER 5

Advanced Ceramics for Microwave Absorber Applications

N. Suresh Kumar[1], K. Chandra Babu Naidu[2,*], Prasun Banerjee[2], H. Manjunatha[3], A. Ratnamala[3] and Sannapaneni Janardan[3]

[1] *Department of Physics, JNTUA, Anantapuramu-515002, A.P, India*

[2] *Dept. of Physics, GITAM Deemed to be University, Bangalore-562163, Karnataka, India*

[3] *Dept. of Chemistry, GITAM Deemed to be University, Bangalore-562163, Karnataka, India*

Abstract: In the present world, electromagnetic radiation pollution has become a problematic issue in order to run electronic equipment as well. At this juncture, the necessity of absorbing the electromagnetic radiation emerged as a significant task. Thus, the advanced ceramic materials for microwave absorption property were developed. In the current chapter, the discussion on the effect of electromagnetic radiation on electronic goods, human health and defense system was elaborated. Furthermore, various ceramic materials were introduced for reducing the electromagnetic radiation pollution thereby microwave absorption process. Moreover, in order to evaluate the microwave absorption property of ceramics, we extracted the parameters like magnetic loss and dielectric loss for each ceramic material. Subsequently, the applications of microwave absorbers in various fields were elucidated.

Keywords: Dielectric loss, Magnetic loss, Magneto-ceramics, Microwave absorbers.

5.1. INTRODUCTION

It was observed that electromagnetic interference (EMI) or electromagnetic (EM) radiation pollution became a severe problem to the electronic goods, human body, military system and other living organisms [1 - 3]. In order to diminish the EMI, mechanisms like electromagnetic reflection, absorption and multiple reflections were developed. These mechanisms were observed consistently, in case of different pure ceramics (bulk & nano), ceramic composites, polymer composites and ceramic polymer composites. Within the three EMI removal mechanisms, the

* **Corresponding author Dr. K. Chandra Babu Naidu:** GITAM Deemed To Be University-Bangalore Campus, Bangalore-562163, Karnataka, India; E-mail: chandrababu954@gmail.com

K. Chandra Babu Naidu & N. Suresh Kumar (Eds.)

microwave absorption mechanism acquired good attention owing to the availability of more number of microwave absorbers materials. Normally, the microwave absorption efficiency depends on the absorption shielding efficiency (SE_A) and is given by $SE_A = 12.3t\ (\omega\sigma\mu_r)^{0.5}$, where ω = the angular frequency, t = the thickness of shielding material and μ_r = relative magnetic permeability [2]. The mathematical equation of SE_A indicated a fact that microwave absorption is dependent on electrical conductivity and magnetic permeability. That is, the microwave absorption property will be increased with an increase in electrical conductivity and magnetic permeability. However, it was a known fact that the electrical conductivity can be induced due to the high oscillating frequency of the spinning electric/magnetic dipoles. This high oscillating frequency of electric and magnetic dipoles can induce dielectric and magnetic loss parameters. Hence, it was confirmed that for achieving the efficient absorption of EM radiation, the absorber materials must show the high magnitude of the magnetic permeability, magnetic loss, dielectric constant and dielectric loss.

Apart from these electromagnetic parameters, the high abundance of physical and morphological parameters such as low density, porous structure and multiple polarization centers can enhance the microwave absorption property. The high magnitude of the above said parameters can give rise to the multiple reflection and scattering of microwaves thereby matching the impedance [1, 3]. From the vast literature survey, it was evidenced that the ceramic materials belong to the spinel ferrites, magnetic spinels, hexaferrites, magnetic perovskites, ceramic composites and ceramic polymer composites showed the high microwave absorption efficiency [2]. In view of this, the nickel zinc ferrites, nickel zinc cobalt ferrites, nickel cobalt zinc lanthanum ferrites, nickel zinc lanthanum ferrites, manganese zinc ferrites, barium zinc cobalt U-hexaferrites, barium hexaferrites, aluminium titanate and lanthanum iron oxide offered considerable microwave absorption behavior at giga hertz frequency range [2]. Once the microwave absorber materials are exposed to the EM radiation, the EM radiation enters into the interior of the material. Later on, the EM energy will be absorbed and converted into dissipated energy. Therefore, the electronic object will be prevented from EMI. A photograph of microwave absorber is provided in Fig. (**5.1**).

Fig. (5.1). Photograph of microwave absorber.

5.2. CERAMIC MATERIALS AND MICROWAVE ABSORBER PARAMETERS

In order to understand the microwave absorption property of ceramics and their composite materials, the electromagnetic parameters, absorption shielding efficiency and reflection loss were recorded in the Table **5.1**. It was seen that some magneto-ceramics and multiferroic ceramics offered high reflection loss (RL) and absorption efficiency (SE_A) at gigahertz frequency in their pure form. For example, the $Ni_{0.4}Co_{0.2}Zn_{0.4}Fe_2O_4$ [10]: -48.1 dB (1.1 GHz), $MnFe_2O_4$ [10]: -42.5 dB (9.4 GHz), $Co_{0.5}Mn_{0.5}Fe_2O_4$ [10]: -47.0 dB (10.5 GHz), $Ni_{0.4}Co_{0.6}BaTiFe_{10}O_{19}$ [10]: -46.9 dB (13.3 GHz), $Ni_{0.2}Co_{0.8}BaTiFe_{10}O_{19}$ [10]: -48.3 dB (16.0 GHz), $(MnNi)_{0.2}Co_{0.6}BaTiFe_{10}O_{19}$ [10]: -53.0 dB (13.5 GHz), $(MnNi)_{0.25}Co_{0.5}BaTiFe_{10}O_{19}$ [10]: -68.9 dB (14.1 GHz) and $La_{0.6}Sr_{0.4}MnO_3$ [10]: -40.9 dB (8.2 GHz) performed the high reflection loss at gigahertz frequency. Due to high RL value, the microwave absorption property of the above-mentioned ceramic material compositions was increased. Hence, we can use these pure materials as the electromagnetic shields/microwave absorbers. In addition to this, all these compositions showed high magnetic loss (at gigahertz) indicating the microwave absorption property. Moreover, the other electromagnetic parameters like dielectric constant, dielectric loss and magnetic permeability also revealed high magnitude of numerical values in support to the microwave absorption nature. On the other hand, the ceramic materials in alloys form like Fe/TiO_2 [10]: -45.1 dB (13.9 GHz), $Co_{1-x}S$ [10]: -45.8 dB (14.0 GHz), FeP [10]: -36.5 dB (15.1 GHz), Co_2P [10]: -23.3 dB (16.7 GHz), Fe/G [10]: -34.3 dB (17.1 GHz), NiCu [10]: -31.1 dB (14.2 GHz), Co_3Fe_7 [10]: -53.6 dB (14.2 GHz), FeSiAl [10]: -39.7 dB (1.4 GHz) and NdFe [10]: -55.9 dB (3.6 GHz) performed the high RL values at GHz frequencies. In comparison with the pure ceramic materials, the RL values of alloyed ceramics were noticed to be little bit smaller in magnitude. This was attributed to the formed low defective structure in the source material upon adding the external element. Even this kind of approach can alter the intrinsic properties of resultant alloyed ceramic material. Therefore, this can give rise to the reduction in the electromagnetic parameters. In the above quoted alloys, several entities showed low magnetic loss values. But interestingly, the alloys like Fe/G (-0.03) and NiCu (-0.02) offered negative value of magnetic loss which can be an indication for the enhancement of microwave absorption property [11, 12]. In the same way, the moderate values of other electromagnetic parameters were observed. Likewise, the ceramic materials in composite form such as Fe_3O_4@CNP [4]: -65.5 dB (9.80 GHz), YS-Fe_3O_4@C [4]: -71.3 dB (14.6 GHz), rGO/Fe_3O_4@SiO_2/NiO [4]: -51.5 dB (14.60 GHz), SiC/SiO_2 [6]: -52.0 dB (10.8 GHz), SiCN nanowires [8]: -53.1 dB (18.0 GHz) and $CoFe_2O_4$:SnO_2/rGO [10]: -54.4 dB (16.5 GHz) showed the high reflection loss values at the GHz frequencies. As we compared with the pure ceramics and alloyed ceramics, the composite ceramics expressed the high RL values suggesting the high microwave absorption behavior at GHz range. But in case of composites, the materials were observed to be in nano-form. The different electromagnetic parameters of composites were displayed in (Table **5.1**). In references [13 - 61], various ferrites, titanates, composites and their microwave absorption property was highlighted. However, all the nanoceramics composites ensured the high porous structure indicating the high microwave absorption efficiency. Therefore, in the present days, the composites have been most commonly used for absorbing the electromagnetic radiation. In this concern, we have different applications of microwave absorbers materials in various fields like industrial, medical and defense.

Table 5.1. Electromagnetic parameters of various ceramic materials.

Ceramic Material	Synthesis Method	Bulk/nano	ε'	ε''	μ'	μ''	SEA (dB)	RL (dB) (GHz)
Hollow Fe_3O_4 [4]	plasma dynamic	Bulk	-	-	-	-	50	-43.9 (10.6)
Fe_3O_4@CNS [4]	-	Nano	-	-	-	-	20	-41.2 (11.54)
Fe_3O_4@CNP [4]	-	Nano	-	-	-	-	40	-65.5 (9.80)
Fe_3O_4/rGO [4]	-	Nano	-	-	-	-	30	-22.7 (12.00)
Ag@Fe_3O_4/rGO [4]	-	Nano	-	-	-	-	50	-40.05 (11.9)
rGO/BaFe12O19/Fe_3O_4 [4]	-	Nano	-	-	-	-	30	-46.04 (15.60)
PANI@MoS_2@Fe_3O_4 [4]	-	Nano	-	-	-	-	30	-49.7 (16.90)
Fe_3O_4/SiO_2 [4]	-	Nano	-	-	-	-	40	-28.6 (8.10)

Ceramic Material	Synthesis Method	Bulk/nano	ε'	ε"	μ'	μ"	SEA (dB)	RL (dB) (GHz)
Fe$_3$O$_4$@SiO$_2$@PPy [4]	-	Nano	-	-	-	-	15	-40.09 (6.00)
Fe$_3$O$_4$@SiO$_2$@rGO [4]	-	Nano	-	-	-	-	20	-26.9 (9.70)
Fe$_3$O$_4$/PPy/CNT [4]	-	Nano	-	-	-	-	20	-25.9 (10.20)
GN/Fe$_3$O$_4$/SiO$_2$/PANI [4]	-	Nano	-	-	-	-	25	-40.7 (12.50)
Fe$_3$O$_4$/PPy/PANI [4]	-	Nano	-	-	-	-	20	-47.3 (13.45)
rGO/Fe$_3$O$_4$ [4]	-	Nano	-	-	-	-	50	-45.0 (8.96)
Fe$_3$O$_4$/C [4]	-	Nano	-	-	-	-	60	-46.0 (15.50)
YS-Fe$_3$O$_4$@C [4]	-	Nano	-	-	-	-	70	-71.3 (14.6)
Fe$_3$O$_4$@SnO2/rGO [4]	-	Nano	-	-	-	-	50	-45.5 (6.40)
rGO/Fe$_3$O$_4$@SiO$_2$/NiO [4]	-	Nano	-	-	-	-	25	-51.5 (14.60)
Ca5Ni4(VO4)6 [5]	SSR	Bulk	-	-	-	-	-	-
SiC/SiO$_2$ [6]	Nanocasting & cold-pressing	Nano	-	-	-	-	-	-52.0 (10.8)
Si$_3$N$_4$@SiCNFs [7]	Gel casting	Nano	-	-	-	-	-	-21.26 (11.35)
SiCN nanowires [8]	Electrospinning	Nano	-	-	-	-	-	-53.1 (18.0)
NiZnFe$_2$O$_4$ [9]	Solution combustion	Nano	-	-	-	-	-	-
Raw CNTs [10]	-	Nano	6.4	1.6	0.98	-0.01	-	-9.9 (10.6)
Purified CNTs [10]	-	Nano	6.5	2.3	0.99	-0.03	-	-14.9 (10.3)
CNT [10]	-	Nano	32.4	33	1	0.01	-	-21.5 (11.4)
Co/CNTs [10]	-	Nano	22.9	9.9	1.1	0.08	-	-22.9 (12.2)
Fe/CNTs (D) [10]	-	Nano	28.6	43	2.02	1.6	-	-18.2 (11.8)
Fe/CNTs (E) [10]	-	Nano	29.7	41	1.94	1.91	-	-24.8 (10.9)
FeCo/CNTs [10]	-	Nano	-	-	-	-	-	-15.5 (9.5)
FeCoNi/CNTs [10]	-	Nano	-	-	-	-	-	-28.2 (15.2)
CNTs [10]	-	Nano	33.2	34.9	1	0.01	-	-22.9 (11.4)
Fe/CNTs [10]	-	Nano	20.2	25.3	1.1	0.1	-	-31.8 (13.2)
FeCNTs [10]	-	Nano	34	55	0.71	0.03	-	-22.7 (15.6)
CoFe$_2$O$_4$/CNTs [10]	-	Nano	-	-	-	-	-	-
Fe$_3$O$_4$/BNCNTs [10]	-	Nano	3.8	0.24	0.26	0.26	-	-18.0 (9.0)
Fe$_3$O$_4$/BNCNTs [10]	-	Nano	3.5	1.9	1.13	0.23	-	-47.9 (10.5)
Ag/MWCNTs 10 wt% [10]	-	Nano	42.5	38.8	0.82	0.01	-	-19.2 (7.8)
Ni/MWCNTs 5 wt% [10]	-	Nano	5.26	2.84	1.11	0.14	-	-23.1 (8.0)

(Table 5.1) cont.....

Ceramic Material	Synthesis Method	Bulk/nano	ε'	ε''	μ'	μ''	SEA (dB)	RL (dB) (GHz)
Ni/MWCNTs -10 wt% [10]	-	Nano	6.02	3.31	1.15	0.22	-	-17.8 (7.2)
Fe/Fe$_3$C/MWCNT [10]	-	Nano	14.9	3.2	1	0.38	-	-12.5 (9.0)
Fe:Fe$_3$C/MWCNT [10]	-	Nano	16	3.9	1.37	0.55	-	-14.1 (4.6)
Er$_2$O$_3$/MWCNT [10]	-	Nano	14.8	4.8	0.99	-0.01	-	-27.7 (10.0)
Sm$_2$O$_3$/MWCNT [10]	-	Nano	10	3.4	0.89	0.02	-	-21.5 (9.4)
CoO$_x$/CF [10]	-	Nano	10.8	1.1	0.5	1	-	-45.2 (13.4)
Fe/CFs (8 wt%) [10]	-	Nano	4.9	1.7	1.5	0.75	-	-41.7 (13.4)
Carbonyl Fe/CF [10]	-	Nano	22	1.05	2.05	1.67	-	-22.9 (9.7)
Fe$_3$O$_4$/rGO [10]	-	Nano	12.3	1.9	1.14	0.16	-	-26.4 (5.3)
NiFe$_2$O$_4$/rGO [10]	-	Nano	8	3	1	0.1	-	-39.6 (9.2)
MnFe$_2$O$_4$/rGO [10]	-	Nano	6.4	3.2	1.22	0.1	-	-29.0 (9.2)
CoFe$_2$O$_4$/rGO [10]	-	Nano	7.1	2.8	1	0.1	-	-37.2 (11.6)
CoFe$_2$O$_4$/rGO [10]	-	Nano	9.4	9.3	0.98	0.5	-	-44.1 (14.7)
CoFe$_2$O$_4$:SnO2/rGO [10]	-	Nano	7.3	2.7	1	0.1	-	-54.4 (16.5)
Fe$_3$O$_4$/C [10]	-	Nano	4.5	2.5	1.04	-0.02	-	-39.3 (15.5)
Fe:SiO$_2$/C [10]	-	Nano	3.5	1	1.14	0.6	-	-20.5 (16.6)
Fe$_3$O$_4$/C [10]	-	Nano	11.4	2.9	1.07	0.12	-	-45.0 (6.2)
Ni$_2$O$_3$/C [10]	-	Nano	7.8	5.2	0.79	-0.21	-	-33.5 (15.0)
SiC [10]	Ball milling	Nano	13.9	2.85	1.18	0.19	-	-27.1 (10.5)
Co/SiC [10]	-	Nano	12.2	6.4	1.08	-0.13	-	-24.3 (11.6)
Ni/SiC [10]	-	Nano	16.5	4.2	1.11	0.13	-	-24.9 (11.8)
Ti/SiC [10]	-	Nano	15.6	5.3	1.09	-0.02	-	-40.2 (8.7)
Zn/SiC [10]	-	Nano	15	5.2	1.09	-0.02	-	-43.7 (9.4)
Ni:Co-P/SiC [10]	-	Nano	23.1	5.9	1	0.04	-	-32.4 (6.4)
Fe/SiC [10]	-	Nano	25	5.8	1.15	0.05	-	-46.3 (6.4)
CNT/Ni/SiC [10]	-	Nano	6.5	1.75	1.1	0.1	-	-37.6 (14.5)
Fe$_3$O$_4$ 30 wt% [10]	-	Nano	9.2	0.2	1.5	0.72	-	-29.5 (3.9)
Fe$_3$O$_4$ 70 wt% [10]	-	Nano	12.1	1.2	1.2	0.6	-	-34.9 (3.4)
(Ni$_{0.4}$Cu$_{0.2}$Zn$_{0.4}$)Fe$_2$O$_4$ [10]	-	-	8.1	-0.01	2.1	3.8	-	-35.6 (1.1)
(Ni$_{0.4}$Co$_{0.2}$Zn$_{0.4}$)Fe$_2$O$_4$ [10]	-	-	7.05	0.03	1.56	1.49	-	-48.1 (1.1)
Ni$_{0.8}$Co$_{0.2}$Fe$_2$O$_4$ [10]	-	-	9.6	-0.13	1	0.3	-	-35.5 (11.5)
Ni$_{0.5}$Co$_{0.5}$Fe$_2$O$_4$ [10]	-	-	8.3	-0.25	0.7	0.3	-	-30.6 (11.9)
MnFe$_2$O$_4$ [10]	-	-	1.5	0.35	4.7	0.73	-	-42.5 (9.4)

(Table 5.1) cont.....

Ceramic Material	Synthesis Method	Bulk/nano	ε'	ε''	μ'	μ''	SEA (dB)	RL (dB) (GHz)
$Co_{0.5}Mn_{0.5}Fe_2O_4$ [10]	-	-	3.8	0.36	7.1	1.02	-	-47.0 (10.5)
$CoFe_2O_4$ [10]	-	-	7.5	0.46	8.9	1.42	-	-39.8 (10.7)
$Bi_{0.8}La_{0.2}FeO_3$ [10]	-	-	9.8	1.4	1	0.9	-	-29.7 (11.5)
$PANI/Co_{0.5}Zn_{0.5}Fe_2O_4$ [10]	-	Nano	19.3	13.1	1.7	0.3	-	-39.8 (22.3)
C/Fe_3O_4 [10]	-	Nano	8.4	0.4	1.3	0.4	-	-44.0 (3.92)
ZnO/Fe_3O_4 [10]	-	-	3.4	1.23	2.4	4	-	-12.92 (1.66)
$Ni_{0.4}Co_{0.6}BaTiFe_{10}O_{19}$ [10]	-	-	6.6	0.3	1.2	0.6	-	-46.9 (13.3)
$Ni_{0.2}Co_{0.8}BaTiFe_{10}O_{19}$ [10]	-	-	6.5	0.4	0.6	0.3	-	-48.3 (16.0)
$(MnNi)_{0.2}Co_{0.6}BaTiFe_{10}O_{19}$ [10]	-	-	6.7	0.4	0.8	0.7	-	-53.0 (13.5)
$(MnNi)_{0.25}Co_{0.5}BaTiFe_{10}O_{19}$ [10]	-	-	6.5	0.4	1.2	0.7	-	-68.9 (14.1)
$BaCe_{0.05}Fe_{11.95}O_{19}$ [10]	-	-	1.6	1.4	1.4	0.77	-	-31.5 (11.3)
$NiFe_2O_4/ZnFe_2O_4/SrFe_{12}O_{19}$ (900°C) [10]	-	-	12.4	5.9	1.2	1.7	-	-27.6 (10.2)
$NiFe_2O_4/ZnFe_2O_4/SrFe_{12}O_{19}$ (1200°C) [10]	-	-	13.4	13	1.3	2	-	-29.7 (10.2)
$Nd_{0.2}Co_{0.2}Sr_{0.8}Fe_{11.8}O_{19}$ [10]	-	-	5.76	0.64	0.97	0.36	-	-20.8 (16.0)
$CoZnAl_{0.2}Ce_{0.2}BaFe_{15.6}O_{27}$ [10]	-	-	3.9	0.08	1.1	0.2	-	-30.4 (6.6)
$Zn_{1.5}Co_{0.5}BaFe_{16}O_{27}$ [10]	-	-	7.4	0.1	1.63	0.87	-	-6.6 (9.4)
$BaCo_{0.9}Zn_{1.1}Fe_{16}O_{27}$ [10]	-	-	4.45	0.33	1.04	0.21	-	-33.6 (12.0)
$Ba_{0.8}La_{0.2}Co_{0.9}Zn_{1.1}Fe_{16}O_{27}$ [10]	-	-	4.11	0.35	0.89	0.24	-	-39.7 (12.7)
$(Sb:SnO_2)Zn:Co:Gd/BaFe_{12}O_{27}$ [10]	-	-	8.3	1.3	1.01	0.2	-	-19.4 (14.8)
$(Ba^{0.7}La_{0.2})_4Co_2Fe_{36}O_{60}$ [10]	-	-	19.1	3	1.4	0.63	-	-23.2 (9.8)
$La_{0.6}Sr_{0.4}MnO_3$ [10]	-	-	12.9	0.3	1.6	0.7	-	-40.9 (8.2)
Fe/TiO_2 [10]	-	-	11.1	0.6	0.9	1.7	-	-45.1 (13.9)
MC/TiO_2 [10]	-	-	3.9	0.4	1.05	-0.01	-	-24.8 (16.1)
$BaTiO_3$ [10]	-	-	23.4	2.7	0.94	0	-	-29.6 (12.0)
$Ni@BaTiO_3$ [10]	-	-	15.6	1.3	0.9	0.2	-	-41.9 (10.7)
$Zn:Fe/ZnO$ [10]	-	-	5.3	0.08	0.8	0.4	-	-20.8 (17.9)
Fe_3Al/Al_2O_3 [10]	-	-	11.5	0.9	1.16	0.44	-	-45.4 (9.0)
Fe_3/Al_2O_3 [10]	-	-	7.3	-2.4	0.81	0.54	-	-21.3 (13.3)
$Co_1\text{-}xS$ [10]	-	-	5.3	1.6	0.9	0.3	-	-45.8 (14.0)

(Table 5.1) cont.....

Ceramic Material	Synthesis Method	Bulk/nano	ε'	ε''	μ'	μ''	SEA (dB)	RL (dB) (GHz)
rGO/CoS$_2$ [10]	-	-	9	4.2	1.2	0.1	-	-56.1 (6.7)
FeP [10]	-	-	18.2	5.4	0.85	0.11	-	-36.5 (15.1)
Co$_2$P [10]	-	-	20.3	4.3	1.05	0.11	-	-23.3 (16.7)
PANI/BaM [10]	-	-	8.6	1.8	4.1	2.1	-	-16.7 (14.7)
Ni$_{0.6}$Zn$_{0.4}$Fe$_2$O$_4$/PANI [10]	-	-	6.2	6	0.9	-0.05	-	-41.0 (12.8)
MnO$_2$/PANI [10]	-	-	10.6	5.9	0.9	-0.04	-	-21.0 (13.6)
Carbonyl Fe [10]	-	-	27.1	0.7	3.5	2	-	-31.4 (3.3)
Fe/G [10]	-	-	7.6	4.3	1.19	-0.13	-	-34.3 (17.1)
NiCu [10]	-	-	12.2	4.4	1.13	-0.02	-	-31.1 (14.2)
CoNi [10]	-	-	9.5	2.2	2	0.6	-	-26.0 (8.5)
Co$_3$Fe$_7$ [10]	-	-	9.2	1.4	1.6	0.5	-	-53.6 (14.2)
FeSiAl [10]	-	-	25.4	1.2	3.5	1.9	-	-39.7 (1.4)
NdFe [10]	-	-	21	3	2.2	0.75	-	-55.9 (3.6)
Ni$_{0.4}$Mg$_{0.6}$Fe$_2$O$_4$ [2]	Microwave heating	Bulk	6.42	8	2.52	3.14	12.1	-8.4
Ni$_{0.4}$Mg$_{0.6}$Fe$_2$O$_4$ [2]	Microwave heating	Nano	9.3	2.7	1.3	-0.01	2.2	-8.4
Ni$_{0.2}$Zn$_{0.8}$Fe$_2$O$_4$ [11]	Microwave heating	Bulk	6.3	5.8	0.88	0.81	6.96	-8.4

5.3. APPLICATIONS OF MICROWAVE ABSORBERS

In the industries, the castle microwave contains the broad range of components offering applications from high performance armoured cable assemblies to radio frequency (RF) and microwave components suitable for harsh environments (see Fig. **5.2**). These components include the microwave absorbers in order to protect all the components from electromagnetic radiation. The microwaves and microwave absorber materials were used in desiccation of tissue without any damage and bleeding. Similarly, the large tumors, liver tumors and lung tumors will be removed without any pain and bleeding. In defense system, the microwave absorbers are used for coating, on the defense equipment, air-crafts, war ships, army cloths, guard front parts, *etc.* As a result of these coatings on different objects of defense system, we can protect the weapons and army people from stray EM radiation. The absorber coating on the air-crafts is shown in Fig. (**5.3**) using dark paint. The castle microwave is one of the leading technologies developing the radio frequency and microwave products. It was a known fact that these castle microwave set up achieved the wide range of applications in the military, space, communication and smart energy systems. Nowadays, the wireless mesh network working at 2.4 and 5 GHz bands provide the solutions for infrastructure networks deployed over different geographical places. Therefore, the area wide security and safety of electronic goods, lighting and traffic light control and data throughput applications can be obtained as well. In addition, the 3D flocked carbon fiber microwave absorbers offered a chance to develop the thin and light weight materials having an ability of high magnitude of EM wave or RF wave suppression excluding the addition of any mass to the electronic design. In other words, these materials are treated as thin shielding materials which showed applications in markets, aerospace, defense, automotive electronics, communication system *etc* [11, 12]. The materials mentioned in Table **5.1** are the cost-effective enough and competitive in all the above mentioned fields.

Fig. (5.2). Photograph of castle microwave components.

Fig. (5.3). Photograph of usage of microwave absorber in defense system.

CONCLUSIONS

The different ceramic materials including their electromagnetic parameters were discussed. It was found from the analysis that the ceramic composites and pure ferrites performed the high electromagnetic absorption property. Furthermore, these materials revealed the applications in defense systems, microwave communications and biomedical fields. Specifically, the ceramic composite materials were considered to be more suitable for microwave device applications due to their efficient electromagnetic parameters.

CONSENT FOR PUBLICATION

Not applicable.

CONFLICT OF INTEREST

The authors confirm that this chapter content has no conflict of interest.

ACKNOWLEDGEMENTS

The authors express thankfulness to Dr. P. Sreeramulu, Assistant Professor (English), GITAM, Bangalore for providing English language editing services to this manuscript.

REFERENCES

[1]   S. Wang, Y. Xu, R. Fu, H. Zhu, Q. Jiao, and T. Feng, "Rational construction of hierarchically porous Fe–Co/N-doped carbon/rGO composites for broadband microwave absorption", *Nano-Micro Lett.,* vol. 11, p. 76, 2019.
[http://dx.doi.org/10.1007/s40820-019-0307-8]

[2]   K.C.B. Naidu, and W. Madhuri, "Microwave Processed Bulk and Nano NiMg Ferrites: A Comparative Study on X-band Electromagnetic Interference Shielding Properties", *Mater. Chem. Phys.,* vol. 187, pp. 164-176, 2017.
[http://dx.doi.org/10.1016/j.matchemphys.2016.11.062]

[3]   D. Zhang, T. Liu, J. Cheng, Q. Cao, G. Zheng, and S. Liang, "… Cao, M.-S. Lightweight and High-Performance Microwave Absorber Based on 2D WS2–RGO Heterostructures", *Nano-Micro Lett.,* vol. 11, p. 38, 2019.
[http://dx.doi.org/10.1007/s40820-019-0270-4]

[4]   L.L. Adebayo, H. Soleimani, N. Yahya, Z. Abbas, F.A. Wahaab, A.T. Ridwan, and H. Ali, "Recent advances in the development of Fe_3O_4-based microwave absorbing materials", *Ceram. Int.,* vol. 46, pp. 1249-1268, 2019.
[http://dx.doi.org/10.1016/j.ceramint.2019.09.209]

[5]   G. Yao, C. Pei, P. Liu, J. Zhou, and H. Zhang, "Low temperature sintering and microwave dielectric properties of Ca5Ni4(VO4)6 ceramics", *Ceram. Int.,* vol. 43, pp. S334-S338, 2017.
[http://dx.doi.org/10.1016/j.ceramint.2017.05.314]

[6]   X. Yuan, L. Cheng, and L. Zhang, "Electromagnetic wave absorbing properties of SiC/SiO$_2$

composites with ordered inter-filled structure", *J. Alloys Compd.,* vol. 680, pp. 604-611, 2016.
[http://dx.doi.org/10.1016/j.jallcom.2016.03.309]

[7]   W. Zhou, L. Long, G. Bu, and Y. Li, "Mechanical and microwave-absorption properties of Si_3N_4 ceramic with SiCNFs fillers", *Adv. Eng. Mater.,* vol. 21, p. 1800665, 2018.
[http://dx.doi.org/10.1002/adem.201800665]

[8]   P. Wang, L. Cheng, and L. Zhang, "Lightweight, flexible SiCN ceramic nanowires applied as effective microwave absorbers in high frequency", *Chem. Eng. J.,* vol. 338, pp. 248-260, 2018.
[http://dx.doi.org/10.1016/j.cej.2017.12.008]

[9]   M. Emília dos Santos, R.B. Kasalb, N.C. Moraesa, G.B. Mota de Meloc, and J.C. Araujo dos Santosa, "André Ben-Hur da Silva Figueiredo, Nanoparticles of $Ni_{1-x}Zn_xFe_2O_4$ used as Microwave Absorbers in the X-band"., *Mater. Res.,* vol. 22, p. e20190188, 2019.
[http://dx.doi.org/10.1590/1980-5373-mr-2019-0188]

[10]   M. Green, and X. Chen, "Recent progress of nanomaterials for microwave absorption", *Journal of Materiomics,* vol. 5, pp. 503-541, 2019.
[http://dx.doi.org/10.1016/j.jmat.2019.07.003]

[11]   K.C.B. Naidu, S.R. Kiran, and W. Madhuri, "Microwave Processed NiMgZn Ferrites for Electromagnetic Interference Shielding Applications", *IEEE Trans. Magn.,* vol. 53, p. 2900207, 2017.

[12]   K.C.B. Naidu, S.R. Kiran, and W. Madhuri, "Investigations on transport, impedance and electromagnetic interference shielding properties of microwave processed NIMG ferrites", *Mater. Res. Bull.,* vol. 89, pp. 125-138, 2017.
[http://dx.doi.org/10.1016/j.materresbull.2017.01.015]

[13]   W. Peng, C. Laifei, Z. Yani, and Z. Litong, "Synthesis of SiC nanofibers with superior electromagnetic wave absorption performance by electrospinning", *J. Alloys Compd.,* vol. 716, pp. 306-320, 2017.
[http://dx.doi.org/10.1016/j.jallcom.2017.05.059]

[14]   Z. Wuqiang, B. Shaowei, C. Haichao, L. Yao, and J. Jianjun, "Electromagnetic and microwave absorption properties of carbonyl iron/MnO_2 composite", *J. Magn. Magn. Mater.,* vol. 358-359, pp. 1-4, 2014.
[http://dx.doi.org/10.1016/j.jmmm.2014.01.033]

[15]   M. Green, A.T. Van Tran, R. Smedley, A. Roach, J. Murowchick, and X. Chen, "Microwave Absorption of Magnesium/Hydrogen-Treated Titanium Dioxide Nanoparticles, Nano", *Mater. Sci.,* vol. 1, pp. 48-59, 2019.

[16]   A. Mitra, A.S. Mahapatra, A. Mallick, and P.K. Chakrabarti, "Enhanced microwave absorption and magnetic phase transitions of nanoparticles of multiferroic $LaFeO_3$ incorporated in multiwalled carbon nanotubes (MWCNTs)", *J. Magn. Magn. Mater.,* vol. 435, pp. 117-125, 2017.
[http://dx.doi.org/10.1016/j.jmmm.2017.03.066]

[17]   A. Borrell, and M.D. Salvador, *Advanced Ceramic Materials Sintered by Microwave Technology.* Sintering Technology - Method and Application, 2018.
[http://dx.doi.org/10.5772/intechopen.78831]

[18]   P. Yadav, S. Rattan, A. Tripathi, and S. Kumar, "Tailoring of complex permittivity, permeability, and microwave-absorbing properties of $CoFe_2O_4$/NG/PMMA nanocomposites through swift heavy ions irradiation", *Ceram. Int.,* vol. 46, pp. 317-324, 2020.
[http://dx.doi.org/10.1016/j.ceramint.2019.08.265]

[19]   Y. Wang, X. Gao, X. Wu, and C. Luo, "Facile synthesis of Mn_3O_4 hollow polyhedron wrapped by multiwalled carbon nanotubes as a high-efficiency microwave absorber", *Ceram. Int.,* vol. 46, pp. 1560-1568, 2020.
[http://dx.doi.org/10.1016/j.ceramint.2019.09.124]

[20]   X. Chu, L. Gan, S. Ren, J. Wang, Z. Ma, J. Jiang, and T. Zhang, "Low-loss and temperature-stable $(1-x)Li_2TiO_3-xLi_3Mg_2NbO_6$ microwave dielectric ceramics", *Ceram. Int.,* vol. 46, pp. 8413-8419,

2020.
[http://dx.doi.org/10.1016/j.ceramint.2019.12.075]

[21]   S. Dong, X. Zhang, D. Zhang, B. Sun, L. Yan, and X. Luo, "Strong effect of atmosphere on the microstructure and microwave absorption properties of porous SiC ceramics", *J. Eur. Ceram. Soc.*, vol. 38, pp. 29-39, 2018.
[http://dx.doi.org/10.1016/j.jeurceramsoc.2017.07.034]

[22]   X. Su, Y. Jia, J. Wang, J. Xu, X. He, C. Fu, and S. Liu, "Preparation and microwave absorption properties of Fe-doped SiC powder obtained by combustion synthesis", *Ceram. Int.*, vol. 39, pp. 3651-3656, 2013.
[http://dx.doi.org/10.1016/j.ceramint.2012.10.194]

[23]   Q. Li, X. Yin, W. Duan, B. Hao, L. Kong, and X. Liu, "Dielectric and microwave absorption properties of polymer derived SiCN ceramics annealed in N_2 atmosphere", *J. Eur. Ceram. Soc.*, vol. 34, pp. 589-598, 2014.
[http://dx.doi.org/10.1016/j.jeurceramsoc.2013.08.042]

[24]   W. Duan, X. Yin, C. Luo, J. Kong, F. Ye, and H. Pan, "Microwave-absorption properties of SiOC ceramics derived from novel hyperbranched ferrocene-containing polysiloxane", *J. Eur. Ceram. Soc.*, vol. 37, pp. 2021-2030, 2017.
[http://dx.doi.org/10.1016/j.jeurceramsoc.2016.12.038]

[25]   W. Hong, S. Dong, P. Hu, X. Luo, and S. Du, "*In situ* growth of one-dimensional nanowires on porous PDC-SiC/Si_3N_4 ceramics with excellent microwave absorption properties", *Ceram. Int.*, vol. 43, pp. 14301-14308, 2017.
[http://dx.doi.org/10.1016/j.ceramint.2017.07.182]

[26]   P.N. Dhruv, S.S. Meena, R.C. Pullar, F.E. Carvalho, R.B. Jotania, and P. Bhatt, "... Basak, C. B. Investigation of structural, magnetic and dielectric properties of gallium substituted Z-type Sr_3Co_2-$GaFe_{24}O_{41}$ hexaferrites for microwave absorbers", *J. Alloys Compd.*, vol. 822, p. 153470, 2020.
[http://dx.doi.org/10.1016/j.jallcom.2019.153470]

[27]   H.B. Bafrooei, B. Liu, W. Su, and K.X. Song, "$Ca_3MgSi_2O_8$: Novel low-permittivity microwave dielectric ceramics for 5G application", *Mater. Lett.*, vol. 263, p. 127248, 2020.
[http://dx.doi.org/10.1016/j.matlet.2019.127248]

[28]   A. Ling, J. Pan, G. Tan, X. Gu, Y. Lou, and S. Chen, "... Liu, X. Thin and broadband $Ce_2Fe_{17}N_{3-\delta}$/MWCNTs composite absorber with efficient microwave absorption", *J. Alloys Compd.*, vol. 787, pp. 1097-1103, 2019.
[http://dx.doi.org/10.1016/j.jallcom.2019.02.164]

[29]   Y. Chen, S. Ma, X. Li, X. Zhao, X. Cheng, and J. Liu, "Preparation and microwave absorption properties of microsheets $VO_2(M)$", *J. Alloys Compd.*, vol. 791, pp. 307-315, 2019.
[http://dx.doi.org/10.1016/j.jallcom.2019.03.338]

[30]   S. Golchinvafa, and S.M. Masoudpanah, "Magnetic and microwave absorption properties of $FeNi_3$/$NiFe_2O_4$ composites synthesized by solution combustion method", *J. Alloys Compd.*, vol. 787, pp. 390-396, 2019.
[http://dx.doi.org/10.1016/j.jallcom.2019.02.039]

[31]   J. Varghese, S. Gopinath, and M.T. Sebastian, "Effect of glass fillers in $Cu_2ZnNb_2O_8$ ceramics for advanced microwave applications", *Mater. Chem. Phys.*, vol. 137, pp. 811-815, 2013.
[http://dx.doi.org/10.1016/j.matchemphys.2012.10.014]

[32]   B. Zhou, Z-Z. Tong, J. Huang, J-T. Xu, and Z-Q. Fan, "Synthesis and thermal behavior of poly(ε-caprolactone) grafted on multiwalled carbon nanotubes with high grafting degrees", *Mater. Chem. Phys.*, vol. 137, pp. 1053-1061, 2013.
[http://dx.doi.org/10.1016/j.matchemphys.2012.11.027]

[33]   F. Ren, X. Yin, R. Mo, F. Ye, L. Zhang, and L. Cheng, "Hierarchical carbon nanowires network modified PDCs-SiCN with improved microwave absorption performance", *Ceram. Int.*, vol. 45, pp.

14238-14248, 2019.
[http://dx.doi.org/10.1016/j.ceramint.2019.04.132]

[34]   S. Dong, C. Lin, and X. Meng, "One-pot synthesis and microwave absorbing properties of ultrathin $SrFe_{12}O_{19}$ nanosheets", *J. Alloys Compd.,* vol. 783, pp. 779-784, 2019.
[http://dx.doi.org/10.1016/j.jallcom.2018.12.265]

[35]   Q. Li, X. Yin, L. Zhang, and L. Cheng, "Effects of SiC fibers on microwave absorption and electromagnetic interference shielding properties of SiC/SiCN composites", *Ceram. Int.,* vol. 42, pp. 19237-19244, 2016.
[http://dx.doi.org/10.1016/j.ceramint.2016.09.089]

[36]   X. Su, J. Wang, X. Zhang, S. Huo, W. Chen, W. Dai, and B. Zhang, "Design of controlled-morphology $NiCo_2O_4$ with tunable and excellent microwave absorption performance", *Ceram. Int.,* vol. 46, pp. 7833-7841, 2020.
[http://dx.doi.org/10.1016/j.ceramint.2019.12.002]

[37]   S.N.A. Rusly, I. Ismail, K.A. Matori, Z. Abbas, A.H. Shaari, Z. Awang, and I.H. Hasan, "Influence of different BFO filler content on microwave absorption performances in $BiFeO_3$/epoxy resin composites", *Ceram. Int.,* vol. 46, pp. 737-746, 2020.
[http://dx.doi.org/10.1016/j.ceramint.2019.09.027]

[38]   P.T. Tho, C.T.A. Xuan, T.N. Bach, D.N.H. Nam, P.M. Tan, and L.T. Ha, "Microwave absorption properties of $La_{1.5}Sr_{0.5}NiO_4$-based nanocomposites", *J. Alloys Compd.,* vol. 805, pp. 1231-1236, 2019.
[http://dx.doi.org/10.1016/j.jallcom.2019.07.212]

[39]   G. Shao, J. Liang, W. Zhao, B. Zhao, W. Liu, and H. Wang, "… Zhang, R. Co decorated polymer-derived SiCN ceramic aerogel composites with ultrabroad microwave absorption performance", *J. Alloys Compd.,* vol. 813, p. 152007, 2019.
[http://dx.doi.org/10.1016/j.jallcom.2019.152007]

[40]   B. Wei, J. Zhou, Z. Yao, A.A. Haidry, X. Guo, and H. Lin, "… Chen, W. The effect of Ag nanoparticles content on dielectric and microwave absorption properties of β-SiC", *Ceram. Int.,* vol. 46, pp. 5788-5798, 2020.
[http://dx.doi.org/10.1016/j.ceramint.2019.11.029]

[41]   L. Zhou, G. Su, H. Wang, J. Huang, Y. Guo, Z. Li, and X. Su, "Influence of NiCrAlY content on dielectric and microwave absorption properties of $NiCrAlY/Al_2O_3$ composite coatings", *J. Alloys Compd.,* vol. 777, pp. 478-484, 2019.
[http://dx.doi.org/10.1016/j.jallcom.2018.10.392]

[42]   F. Khan, B. Amatya, J.F. Pallant, I. Rajapaksa, and C. Brand, "Multidisciplinary rehabilitation in women following breast cancer treatment: A randomized controlled trial", *J. Rehabil. Med.,* vol. 44, no. 9, pp. 788-794, 2012.
[http://dx.doi.org/10.2340/16501977-1020] [PMID: 22858869]

[43]   Z. Yang, F. Luo, Y. Hu, D. Zhu, and W. Zhou, "Dielectric and microwave absorption properties of TiAlCo ceramic fabricated by atmospheric plasma spraying", *Ceram. Int.,* vol. 42, pp. 8525-8530, 2016.
[http://dx.doi.org/10.1016/j.ceramint.2016.02.078]

[44]   X. Long, C. Shao, and J. Wang, "Continuous SiCN fibers with interfacial SiC_xN_y phase as structural materials for electromagnetic absorbing applications", *ACS Appl. Mater. Interfaces,* vol. 11, no. 25, pp. 22885-22894, 2019.
[http://dx.doi.org/10.1021/acsami.9b06819] [PMID: 31198023]

[45]   I. Cacciotti, M. Valentini, M. Raio, and F. Nanni, "Design and development of advanced BaTiO3/MWCNTs/PVDF multi-layered systems for microwave applications", *Compos. Struct.,* vol. 224, p. 111075, 2019.
[http://dx.doi.org/10.1016/j.compstruct.2019.111075]

[46]   G. Feng, W. Zhou, C-H. Wang, Y. Qing, D. Chen, L. Gao, and D. Zhu, "Microwave absorption of M-

type hexaferrite $Ba_{1-x}Ca_xFe_{12}O_{19}$ ($x\leq0.4$) ceramics in 2.6–18 GHz", *Ceram. Int.,* vol. 45, pp. 7102-7107, 2019.
[http://dx.doi.org/10.1016/j.ceramint.2018.12.214]

[47] X. Meng, Q. Han, Y. Sun, and Y. Liu, "Synthesis and microwave absorption properties of $Ni_{0.5}Zn_{0.5}Fe_2O_4/BaFe_{12}O_{19}$@polyaniline composite", *Ceram. Int.,* vol. 45, pp. 2504-2508, 2019.
[http://dx.doi.org/10.1016/j.ceramint.2018.10.179]

[48] M.S. Mustaffa, R.S. Azis, N.H. Abdullah, I. Ismail, and I.R. Ibrahim, "An investigation of microstructural, magnetic and microwave absorption properties of multi-walled carbon nanotubes/$Ni_{0.5}Zn_{0.5}Fe_2O_4$", *Sci. Rep.,* vol. 9, no. 1, p. 15523, 2019.
[http://dx.doi.org/10.1038/s41598-019-52233-2] [PMID: 31664142]

[49] G. Feng, W. Zhou, H. Deng, D. Chen, Y. Qing, C. Wang, and Y. Zhou, "Co substituted $BaFe_{12}O_{19}$ ceramics with enhanced magnetic resonance behavior and microwave absorption properties in 2.6 – 18 GHz", *Ceram. Int.,* vol. 45, pp. 13859-13864, 2019.
[http://dx.doi.org/10.1016/j.ceramint.2019.04.083]

[50] H. Jia, W. Zhou, H. Nan, Y. Qing, F. Luo, and D. Zhu, "High temperature microwave absorbing properties of plasma sprayed $La_{0.6}Sr_{0.4}FeO_{3-\delta}/MgAl_2O_4$ composite ceramic coatings", *Ceram. Int.,* vol. 46, pp. 6168-6173, 2020.
[http://dx.doi.org/10.1016/j.ceramint.2019.11.083]

[51] Y. Dong, X. Yin, H. Wei, M. Li, Z. Hou, H. Xu, and L. Zhang, "Carbon nanowires reinforced porous $SiO_2/3Al_2O_3 \cdot 2SiO_2$ ceramics with tunable electromagnetic absorption properties", *Ceram. Int.,* vol. 45, pp. 11316-11324, 2019.
[http://dx.doi.org/10.1016/j.ceramint.2019.02.209]

[52] E.İ. Şahin, S. Paker, and M. Kartal, "Characterization, Production and Microwave Absorbing Properties of Polyaniline-$NiFe_2O_4$: Tb Composites", *Mater. Sci.,* vol. 25, 2019.
[http://dx.doi.org/10.5755/j01.ms.25.3.20810]

[53] S. Liu, K. Wei, Y. Cheng, B. Qin, S. Yan, H. Luo, and L. Deng, "Enhanced microwave absorbing properties of La-modified $Bi_5Co_{0.5}Fe_{0.5}Ti_3O_{15}$ multiferroics", *J. Mater. Sci. Mater. Electron.,* vol. 30, pp. 15619-15626, 2019.
[http://dx.doi.org/10.1007/s10854-019-01940-7]

[54] Y. Li, H. Cheng, N. Wang, S. Zhou, D. Xie, and T. Li, "Preparation and microwave absorption properties of the $Fe/TiO_2/Al_2O_3$ composites", *Nano,* vol. 13, p. 1850125, 2018.
[http://dx.doi.org/10.1142/S1793292018501254]

[55] Y. Duan, Y. Cui, B. Zhang, G. Ma, and W. Tongmin, "A novel microwave absorber of FeCoNiCuAl high-entropy alloy powders: Adjusting electromagnetic performance by ball milling time and annealing", *J. Alloys Compd.,* vol. 773, pp. 194-201, 2019.
[http://dx.doi.org/10.1016/j.jallcom.2018.09.096]

[56] M. Ning, B. Kuang, Z. Hou, L. Wang, J. Li, Y. Zhao, and H. Jin, "Layer by layer 2D MoS_2/rGO hybrids: An optimized microwave absorber for high-efficient microwave absorption", *Appl. Surf. Sci.,* vol. 470, pp. 899-907, 2019.
[http://dx.doi.org/10.1016/j.apsusc.2018.11.195]

[57] R. Bhattacharyya, Jyoti, Gupta, A., Prakash, O., Roy, S., Bhattacharyya, T. K., Das, S. "*In-situ* synthesis of ($Mg_{0.5}Zn_{0.5}$) Fe_2O_4-graphene oxide nanocomposite for broadband microwave absorption in GHz frequency range", *2019 URSI Asia-Pacific Radio Science Conference (AP-RASC),* 2019.
[http://dx.doi.org/10.23919/URSIAP-RASC.2019.8738396]

[58] M. Rostami, and M.H. Majles Ara, "The dielectric, magnetic and microwave absorption properties of Cu-substituted Mg-Ni spinel ferrite-MWCNT nanocomposites", *Ceram. Int.,* vol. 45, pp. 7606-7613, 2019.
[http://dx.doi.org/10.1016/j.ceramint.2019.01.056]

[59] S. Wang, K. Hu, M. Zhang, P. Zhang, K. Zhang, X. Kong, and Q. Liu, "Magnetic-field-induced

synthesis of one-dimensional core/shell Fe_3O_4/carbon nanorods composites as a highly efficient microwave absorber", *Ferroelectrics,* vol. 548, pp. 220-231, 2019.
[http://dx.doi.org/10.1080/00150193.2019.1592529]

[60] M. Yang, W. Zhou, Y. Liu, L. Li, F. Luo, and D. Zhu, "$LiCo_xNi_{1-x}O_2$ with high dielectric and microwave absorption performance in X-band", *Ceram. Int.,* vol. 45, pp. 17800-17805, 2019.
[http://dx.doi.org/10.1016/j.ceramint.2019.05.351]

[61] Z. Teng, S. Zeng, W. Feng, L. Zhu, Y. Tan, X. Han, and H. Zhang, "Facile synthesis and enhanced microwave absorption properties of Fe-Fe_3C@C composites", *J. Mater. Sci. Mater. Electron.,* vol. 30, pp. 14573-14579, 2019.
[http://dx.doi.org/10.1007/s10854-019-01829-5]

CHAPTER 6

Advanced Ceramics and Their Environmental Applications

N.V. Krishna Prasad[1,*], M.S.S.R.K.N. Sarma[2], B.V. Rama[3], K. Niranjan[2] and K.V. Ramesh[3]

[1] *Department of Physics, GITAM Deemed to be University, Bangalore-562163, Karnataka, India*

[2] *Department of Physics, Andhra University, Visakhapatnam-530003, A.P. India*

[3] *Department of Physics, GITAM Deemed to be University, Visakhapatnam-530045, AP, India*

Abstract: Advanced ceramics are substances that are used for the preparation of ceramic materials with special properties. They are also called fine, engineering, high performance, high tech or technical ceramics. Traditionally, ceramics are known to be inorganic, non-metallic solids made of materials being powdered and then fabricated into material by applying heat which exhibit strength, hardness brittleness and low electrical conductivity. This chapter mainly deals with different types of advanced ceramics, their characteristics and their environmental applications.

Keywords: Environmental Applications, Technical Ceramics.

6.1. INTRODUCTION

Advanced or Technical ceramics come under the category of:

1. Material which is a combination of existing materials that undergo the process of modern approach in order to achieve surprising variations of regular properties that are exhibited by traditional ceramics.

2. Tough as well as electrically conductive ceramic products as some metals.

Processing of advanced ceramics and their development plays a significant role in going at a rapid pace. This is considered to be a revolution in terms of materials and the properties observed. Versailles project on advanced materials and standards (VAMAS) in 1993 has redefined the advanced ceramic as a non-metallic ceramic which is inorganic, which is a basic crystal of highly controlled

---

* **Corresponding author Dr. N.V. Krishna Prasad:** GITAM Deemed to Be University-BLR Campus, Bangalore-562163, Karnataka, India; E-mail: drnvkprasad@gmail.com

composition, with detailed regulation and manufactured from raw materials that are highly refined and characterized to give specified and precise attributes [1].

An advanced ceramic has a number of distinguished features like:

1. Basically crystalline (without glassy component nature).

2. Exhibiting microstructures which are engineered to have the size and shape of a grain with porosity.

3. Having large phase distributions *i.e.* second phases such as fibres and whiskers are arranged, controlled and carefully planned. This process requires processing and composition with perfect regulation. Processing includes clean-room processing as the norm and pure synthetic compounds should be used in place of raw materials that occur naturally which are used as precursors in manufacturing.

4. Having a tendency to exhibit the unique electrical property of superconductivity, mechanical properties like increased toughness and very high strength in temperature. These properties can be carefully processed and can undergo quality control precisely.

5. Having a design with micro structure and micro control in processing.

Hence technical or advanced ceramics are often treated as products with high added values.

Advanced or technical ceramics belong to one of several materials, such as glass, plastics and metals that can be used to protect the environment. These ceramics are also helpful in regenerating damaged ecosystems. The technology used to create these materials is known as environmental engineering [2].

6.2. CHARACTERISTICS OF ADVANCED CERAMICS

Aluminium Oxide (Al_2O_3), Silicon Carbide (SiC), Aluminium Nitride (AlN), Zirconia (ZrO_2), Silicon Nitride ($Si_3N_{4)}$ and materials with titanium base are some advanced ceramics each having particular characteristics of their own. These materials exhibit high performance and act as an alternative in terms of economic parameter to regular materials like plastics, metal and glass. One of the important aspects of recent applications is to enhance low cost operation. In view of this adoption of new materials, the science and engineering meet the needs of unique and specific applications to gain significance. Ceramics can be joined to metals with special expertise. In this context it is required to remember worldwide company which is famous for metallisation and ceramic joining (Morgan advanced materials) that have worldwide customers. Morgan is expert in providing top most integrated solutions for components of any size, shape or specification. Any designer has to take into consideration the types of ceramic materials available. Many of the ceramic materials are highly favourable with excellent track record of electrical, chemical, mechanical and thermal properties. Consideration of physical parameters that include strength, hardness, wear resistance, thermal stability and corrosion resistance is very important while selecting a material. Optimization of above properties can be done based on the selected material. Examples include alumina, aluminium nitride, zirconia, silicon nitride *etc* [3 - 5].

6.3. TYPES OF CERAMICS

a. Alumina

One of the important materials with combination of good electrical and mechanical properties is alumina which is suitable for many applications that include X-ray and electron tubes, aerospace devices, laser devices, flow meters, applications related to high vacuum, pressure sensors and wear components *etc*. Alumina has high strength, resistance, stiffness and hardness. Its grade ranges from 60.0 to 99.90% along with additives designed to increase the dielectric strength. Formation of alumina is done with ceramic processing methods of different types. It is capable of processed into different shapes and sizes. Also, it is capable of getting joined to other metals or ceramics by using special techniques like brazing and metalizing [6].

b. Aluminum Nitride (AlN)

AlN exhibits good thermal conductivity, corrosion resistance and thermal shock resistance and Because of these properties AlN can be widely used in power and opto-electronics, aeronautical and railway systems, processing of semiconductors, defence and microwave applications. Specific applications include IC packages, heat sinks and heaters.

c. Zirconia (Zirconium Dioxide)

Zirconia (zirconium dioxide) is a ceramic which is a white powdered metal oxide. It is made from a metal zirconium having properties similar to titanium. Since it is chemically unreactive zirconium is an important choice for dental applications. The main advantage of zirconia is its capacity of exhibiting resistance at 2400°C (huge temperature) and corrosion. This temperature is greater than the melting point (2072°C) of alumina. The two combinations of zirconia are magnesia partially stabilised zirconia known as (Mg-PSZ and yttrium tetragonal poly-crystal zirconia known as Y-TZP which are nicely suitable for engineering the required hardness, mechanical strengths, wear and corrosion resistance in exceptional cases (structural applications). Fully stabilised zirconia (FSZ) grades and its development for nozzles, crucibles and other components used for molten metal handling applications is the result of high temperature capabilities exhibited by zirconia [7].

d. Silicon Nitride (Si_3N_4)

Silicon Nitride was formed when a high strength and high toughness material is required. Silicon Nitride is widely used in engine components, cutting tools and bearings. It exhibits very high temperature strength, good creep as well as excellent oxidation resistance. It has low coefficient for thermal expansion which can provide excellent thermal shock resistance as compared to other ceramics. Its properties include hardness, fracture toughness of high value, chemical resistance as well as wear resistance. Si_3N_4 is manufactured in three types of products given by RBSN known as Reaction Bonded Silicon Nitride, HPSN known as Hot Pressed Silicon Nitride and SSN known to be Sintered Silicon Nitride. Some of its specific applications to be mentioned are Cutting tools, Bearing ball and roller elements, Valves, Glow plugs, Thermocouple sheaths, Turbocharger rotors for engines welding nozzles, jigs and fixtures [8].

e. Silicon Carbide (SiC)

Silicon carbide (SiC) was also called as carborundum. It is a semiconductor with combination of *silicon* and carbon. It is naturally occurred as a rare mineral moissanite. silicon carbide exhibits temperature strength of high value and thermal resistance up to a temperature of 1650°C. It is highly wearing resistant with high hardness, low density, and excellent chemical resistance [9]. SiC is used in permanent and moving turbine components, bearings, seals, semiconductor wafer processing equipment and ball valve parts. It is also significant in the area of applications related to thermal processing like beams, rollers, batts, plate setters profiled supports and protective sheaths.

6.4. COMPONENTS AND THEIR SHAPE

Shape is an important consideration once we decide any material. It is a known fact that some certain shapes cause weaknesses in the component. In view of this, the simple shapes are selected while working with ceramics as they are observed to provide strong results on continuous basis.

6.5. METALLISING/BRAZING

Connecting of ceramics to metals is required for some applications. Some of the methods for joining include friction welding, adhesive bonding and mechanical fastening. But the extensively method of usage is brazing which creates perfect and hard joint between ceramic & metal without any leaks This process of brazing is a method that starts with formation of layer (metallisation) that creates a wettable surface on the ceramic so that there will be flow of braze alloy between the two components and upon the surface [10].

6.6. CO-FIRED ASSEMBLIES

Co-fired Assemblies denote the monolithic ceramic devices where a ceramic and dielectric are fired in a klin at same time. Some of the co-fired assemblies include resistors, inductors, capacitors, transformers and hybrid circuits. They are also used in specific applications like flow meters where a wire is placed inside during pre-sintered stage of ceramic. Finally, the sintering process shrinks the ceramic and the wire compresses on the metal forming gas tight seal.

6.7. COATING AND GLAZING

Grain size reduces roughness of ceramic while glazing. Large sized grain leads to cavities after grinding. Hence glazing needs to be done to obtain smooth and excellent surface finish.

6.8. ADVANTAGES OF TECHNICAL CERAMICS

Advanced technical ceramics provide a wide range of benefits which have been well established within all levels of industrial production methods and applications related to advanced technology. Latest of their applications are in advanced nuclear technology which control the level of air pollution, used as bio fuel, in carbon sequestration, for coal gasification, in environmental remediation, high energy storage capacity, green building technology and solar technology *etc.*

a. Solid Oxide Fuel Cells (SOFCs)

Since its initial introduction into the market in 2002, SOFCs have demonstrated a promising potential as efficient and pollutant-free energy sources. The consumption of O_2 and fuel, which is in the form of H_2 or hydrocarbons, results in the occurrence of two reactions on each side of membrane (1.Oxidation and 2. Reduction). This chemical energy is then converted into electrical energy as a result of the rapid ionic conduction of either oxygen (O_2) or protons across the solid oxide membrane, which allows for a charge transport to arise. SOFCs use ceramic materials of oxide-ion conducting type as an electrolyte in order to ensure that electrons are released and electricity is produced. SOFCs are unique in that they exhibit high ionic conductivity, density and electrical resistance, all the while also maintaining their stability when exposed to high operating temperatures of up to 500°C. The high efficiency of SOFCs significantly reduces the amount of H_2 or hydrocarbon fuel, thereby ultimately reducing the potential carbon dioxide (CO_2) or nitric oxide (NO_x) emissions associated with this energy option. In fact, SOFCs have been shown to release the lowest amount of greenhouse gas (GHG) emissions as compared to all other energy sources which are non-renewable. Combination with fuel cell technology enhances efficiency of SOFCs by as much as 80%.

6.9. ENVIRONMENTAL APPLICATIONS OF ADVANCED CERAMICS

a. λ (lambda) Sensor

λ (lambda) sensor is an exhaust gas oxygen sensor which signifies an important role in automobile industry. This is used in vehicles for measurement as well as to control ratio of air *versus* fuel in gases that are exhausted. This allows reducing engine emissions and acts as three-way catalytic converter. This results in decrease of air pollution which is very important for surrounding environment. The principle behind the lambda sensor is Nernst principle. It is a galvanic oxygen con-centration cell with solid electrolyte. The solid-state electrolyte consists of a gas impermeable ceramic element made of zirconium dioxide. The ciconoin dioxide is stabilized with yttrium oxide and closed at one end. The inner and outer surfaces of ceramic element are provided with electrodes made of thin porous layer of platinum which has an influence on the sensor characteristic due to its catalytic effects on the one side and on the other side it acts as electrical contact. The outside surface of the ceramic sensor is coated with a platinum layer made of high porous ceramic layer which extend deep into the exhaust gas stream. Hence this strength layer will prevent the platinum (catalytic) layer from corrosive and erosive effects of the deposits in the exhaust gas. Hence the sensor will be highly stable and can perform for long time duration [11].

b. Operation

The Fig. (**6.1**) [12] shows a lambda sensor installed into an engines exhaust system. The sensor is installed at a point such a way that the temperature required for efficient functioning of the sensor is provided for complete operating range of the engine [12]. As already mentioned, the sensor is kept deep into the exhaust gas. It is designed such that the surface of one electrode is surrounded by the exhaust gas while the other end of the electrode is connected to the outer atmosphere. Irrespective of excess fuel conditions still the residual oxygen is found to be $\lambda = 0.95$ in the engines exhaust. This value is approximately equal to 0.1 to 0.3% by volume. By using porous platinum electrodes implies that catalytic conversion of residual oxygen take place with the carbon monoxide, hydrocarbons, and the hydrogen present in the exhaust system. After complete conversion the remaining residual oxygen is a function of the exhaust gas λ (lambda) value which is measured by the oxygen (lambda) sensor. Transition from high residual oxygen content (lean mixture) to very low residual oxygen content (rich mixture) the residue of oxygen content decreases suddenly in the stoichiometric region where the value of λ is equal to unity of air and fuel mixture. This will result in the voltage output of sensor suddenly jumping to $\lambda = 1$. The voltage of the sensor as well as resistance (internal) depends on temperature. Above 350°C in case of unheated sensor and above 150°C in case of heated sensor desired closed-loop control is set for efficient functioning [13].

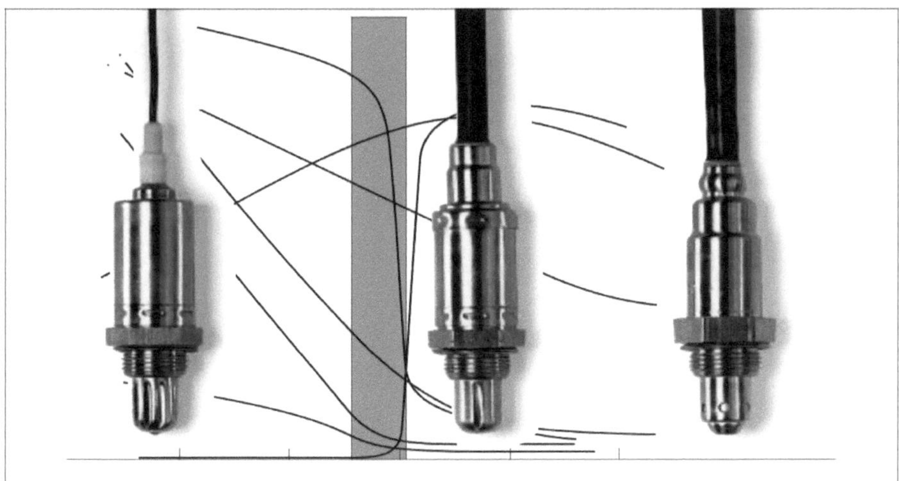

Fig. (6.1). Lambda Sensor (Courtesy: BOSCH).

c. Wind Turbine Technology

It is known fact that wind turbines cause serious corrosion problems Fig. (**6.2**) [14] as they are always exposed moisture, dust, moisture, ultraviolet radiation and temperature change which are unavoidable environmental conditions [14]. To deal this problem a group of researchers belonging to Henkel corporation proposed a solution which leads to an electro ceramic coating called Alodine EC2 in the year 2007 [15]. This electro ceramic coating can protect any difficult environments and hence it can be coated to any component of wind turbine. This coating is found to be environmentally friendly, hard as well as flexible. Another important feature of electro ceramic coating is to provide good resistance, better corrosion, withstand high temperature and also withstand chemical changes for components [16]. In addition, it can replace existing processes of coating which include organic coating, electroplating and heavy zinc phosphate coating. Protection components used in wind turbines that are located onshore & offshore can be attained by using green coating process. This coating is well designed to meet the standards of ELV (End of Life), WEEE (Waste Electrical and Electronic Equipment) and RoHS (Restriction of Hazardous Substances) which do not contain any heavy metal. Apart from the above advantages electro ceramic will provide aesthetic finish and increases the life and service of each component [17].

Fig. (6.2). Environmental Changes leading to corrosion. (Courtesy: Image Credit Photos.com).

d. Photo Voltaic Systems

Advanced ceramics have the capability of withstanding temperatures above 1000°C. Hence, they can be perfectly used in photovoltaic, solar energy conversion as well as for other power plant engine systems. In addition, ceramic components have high resistance to thermal shock, wear and corrosive attack. Hence, they can be used in environmental technology in water processing and treatment industry Fig. **(3)** [18].

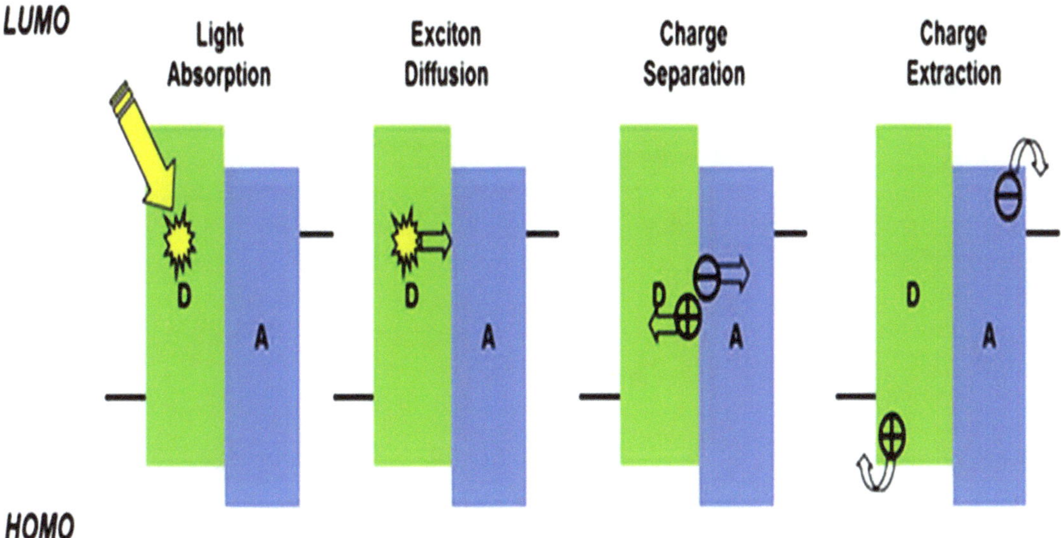

Fig. (6.3). Water Processing (Courtesy: Shutterstock).

e. Waste Sortage

Advanced ceramics can be used in transporting and sorting waste during recycling process because they are corrosion resistive [19]. This property was utilized by morgan technical ceramics. During the process of developing ceramic membranes for the purpose of water filtration as well industrial waste water purification. This company was located in Worcestershire (UK) which manufactures ceramic components that can be used in various industries in order to make them more environmentally friendly [20]. Big number of technical ceramics were readily available for use in the environmental engineering sector. These range from dens sintered monolithic oxide ceramics to filter ceramics, including piezoceramics, magnetic ceramics and fiber reinforced ceramics [21].

f. Clean Technology

Technical ceramic materials and their incorporation into different components of clean technology industry have been continuously increasing for the past few years [22]. While certain clean technology products, such as solid oxide fuel cell (SOFC) systems, are entirely composed of technical ceramics [23], other areas within this industry have found that replacing single critical components, such as insulation materials, with ceramics is equally beneficial.

g. Green Building Design

Since traditional construction methods have been associated with causing adverse impacts to the environment in terms of increase CO_2 and GHG emissions, construction companies have turned to green building technologies to save on resources and reduce their environmental impact. For example, ceramic tile, which is typically comprised of all-natural ingredients including clay, sand, feldspar, quartz and water, is often incorporated into green building projects. In addition to eliminating the need to add any chemicals or volatile compounds into ceramic tile mixtures, many companies will also use recycled material and old ceramic tiles to create new ceramic tiles [24].

h. Solar Energy

It is a known fact that demand for solar energy continues to rise in comparison with conventional power sources in existence. In this context extensive research has been taking place to reduce the cost associated with generation and storage of solar power. It is reported that a team of researchers from Purdue university have replaced the materials that are currently used in solar power plants heat exchangers with a novel ceramic-metal composite of zirconium carbide and tungsten [25]. This significantly reduced the cost of the material as compared to stainless steel or nickel alloy-based heat exchangers and also the customized channels that exist between the elements of this composite have been improved the heat exchange.

i. Piezoceramic Sensors

Sensors with sensing material being both piezoelectric and ceramic are piezoceramic sensors. Lead zirconate titanate commonly called PZT is used as piezoceramic in the sensor. It is a fact that piezoelectric effect exists in various physical environments. These sensors can measure pressure, temperature, acceleration, strain, force or induced shock. Environmental changes that generate a force on this sensor result in an electric charge which is detected and measured. Stress/strain gauges used in civil engineering represents a piezoceramic sensor which picks the displacement of a structure of any size when it experiences force. Majority of them can measure multiple physical parameters. The main aspect of using piezoceramic materials in sensors is their viability for any geometry without change in basic properties [26]. The physical bulk structure doesn't change even they generate an electrical charge under compression which allows them not to break under continuous deformation. Resistant towards electromagnetic radiation enable piezoceramic materials for usage in various environments and wide areas of applications.

j. Automotive Industry

Advanced ceramics which are made with piezo-ceramic components can be used extensively in automotive industry. One main example includes using advanced ceramic as sensor in electronic control unit to provide the information related to respective vehicle engine operation, position of the vehicle as well as directional changes. The electronic components made with ceramic substrates respond to this information for efficient motor management and control the safety systems anti-lock braking, anti slip regulation and also capable of releasing airbags when required. Also, efficiency will be increased in the engine where heat resistant ceramic parts like backings (included in crankshaft housing), valve components water components and fuel pumps. They also ensure less wear and noise emission. These ceramics are also used for construction of light metals. Ceramic components used in xenon, halogen and LED lights improve the visibility in a significant manner. Ceramic (armor) components used in ballistic vehicle protection exhibit safety required for defence and emergency vehicles. SPK ceramic cutting materials and precision tools can be used to ensure reliable and efficient vehicle components which are made from cast iron/hardened steels.

k. Electronics Industry

Advanced ceramics are widely used in electronic industry in the form of many components including substrates, core materials and circuit carriers. Examples include heat sinks to absorb temperature in high-power electronic circuits. One of the leading manufacturers for ceramic components used in automotive industry, aerospace technology, optoelectronics, measurement and control technology as well as consumer electronics is Ceram. Tec.

6.10. AEROSPACE INDUSTRY

Ceramics are primarily used in aerospace industry. They are used in engine, thermal protection shields, and exhaust systems. They are also used as structures for UHS (ultra high speed) flying objects. Hypersonic vehicles that can withstand temperatures upto 2,200°C use ultra high temperature ceramics (UHTCs) for their fabrication. Ceramics are also used in defence aircrafts, space shuttles and other commercial aircrafts. In general, ceramic materials are lighter than metals which make them play vital role in aerospace industry. Light weight is one of the applications in designing concorde (only supersonic airliner in 1990's) which selected a machinable glass ceramic. It is also an electrically insulating material used in the engine control as well as management system. At high temperatures this material is very stable and found to be machined like plastic. This property makes it more attractive for this application. TBC's (thermal barrier coatings) located in the hot part of the engine use structural ceramics. Glass ceramics (machinable) are used at windows, doors and hinge points by U.S. Space Shuttle Orbiter program team and NASA in their aircrafts [27]. Advanced ceramics have become part and parcel of various engine parts for the past 30 to 40 years. The development of silicon carbide (SiC) and its composites gained significance for use in jet engine turbines with main concentration on turbine blades. This will stop the metal alloy blades from melting and the jet engine can be run without any cooling of channels and efficiency of fuel will be increased. Blades made of ceramic composites could withstand temperatures of 1500 to 1600°C and the engine can run at higher temperatures. This increases the energy efficiency leading to less fuel consumption.

CONCLUSIONS

To conclude selection of optimum material depends on various factors. One of the important aspects in consideration of an application and its performance will be dependent on properties like electrical, thermal chemical and mechanical properties. Hardness, heat resistance,, chemical inertness, superior electrical properties, physical stability and biocompatibility decides the suitability of a ceramic material to be used in products that make ceramic materials most important groups of materials in the world.

CONSENT FOR PUBLICATION

Not applicable.

CONFLICT OF INTEREST

The authors confirm that this chapter content has no conflict of interest.

ACKNOWLEDGEMENTS

The authors express thankfulness to Dr. P. Sreeramulu, Assistant Professor (English), GITAM, Bangalore for providing English language editing services to this manuscript.

REFERENCES

[1]   https://www.ceramtec.com/news/archive/year/2014/id/1843/silicon-nitride-ceramics-for-ro-ler-bearing-technology/

[2]   www.ceramicindustry.com

[3]   Treffpunkt Keramik, *Environmental Engineering..* n.d. Retrieved from: https://www.ceramicapplications.com/ applications.html? cat1_id=2195

[4]   J. Olenick, V. Venkateswaran, G. Korbut, and J. Newkirk, *Ceramics in Clean Tech.,* 2016.

[5]   W.R. Matizamhuka, "Advanced ceramics – the new frontier in modern-day technology: Part I", *J. South. Afr. Inst. Min. Metall.,* vol. 118, pp. 757-764, 2018. [REMOVED HYPERLINK FIELD].
[http://dx.doi.org/10.17159/2411-9717/2018/v118n7a9]

[6]   N.K. Patel, S.R. Bishop, R.G. Utter, D. Das, and M. Pecht, "Failure Modes, Mechanisms, Effects and Criticality Analysis of Ceramic Anodes of Solid Oxide Fuel Cells", *Electronics (Basel),* vol. 7, p. 323, 2018.
[http://dx.doi.org/10.3390/electronics7110323]

[7]   Where can advanced ceramics be found?, *Ceram Tec.*https://www.ceramtec.com/ files/ca_where_can_advanced_ceramics_be_found.pdf

[8]   "Green materials report: Ceramic tile", *Green Building Elements.*https://greenbuildingelements.com/2014/07/03/green-materials-report-ceramic-tile/

[9]   "Ceramic metal composite could lower cost of electricity from solar power", *The American Ceramic Society.*https://ceramics.org/ceramic-tech-today/energy-1/ceramic-metal-composite-could-lo-er-cost-of-electricity-from-solar-power

[10]   Benedette Cuffari (Application of Technical Ceramics in Clean Technology), http://www.azocleantech.com/article.aspx

[11]   M. Singh, "Integration Science and Technology of Advanced Ceramics for Energy and Environmental Applications", *Conf. of Brazil-MRS (SBPMat),* 2013 Florianopolis, Brazil

[12]   M.R. Pascucci, and R.N. Katz, *Modern Day Applications of Advanced Ceramics. Inter ceram,* vol. 42, no. 2, pp. 71-78, 1993.

[13]   R. Morrell, Ceramics in Modern Engineering.*Engineering Applications of Ceramic Materials.,* M.M. Schwartz, Ed., American Society for Metals Metals Park OH, 1985, pp. 3-12.

[14]   L.B. Sibley, and M. Zlotnick, "Considerations for Tribological Applications of Engineering Ceramics", *Mater. Sci. Eng.,* vol. 71, pp. 283-293, 1985.
[http://dx.doi.org/10.1016/0025-5416(85)90238-1]

[15]   A. Bennett, "Requirements for Engineering Ceramics in Gas Turbine Engines", *Mater. Sci. Technol.,* vol. 2, no. 9, pp. 895-899, 1986.
[http://dx.doi.org/10.1179/mst.1986.2.9.895]

[16]   A. Krauth, and K. Berroth, Engineering Ceramics for Industrial Applications; Wear-, Heat-, and Automotive Technology.*Engineering Applications of Ceramic Materials.,* M.M. Schwartz, Ed., American Society for Metals Metals Park OH, 1985, pp. 308-313.

[17]   *Coors Tek Internet site.*http://www.coorstek.com/coorstek/

[18]   D. Zeus, "How the Use of Advanced Ceramics as Tribo material Has Affected the Evolution of Mechanical Seals", *Ceram. Forum Int.,* vol. 68, no. 1–2, pp. 36-45, 1991.

[19]   Y. Hara, "Application of Fine Ceramics in Industrial Fields", *Boshoku Gijutsu,* vol. 38, no. 7, pp. 527-537, 1989.

[20]   C. Toy, and T. Baykara, "Yuzyilin Malzemesi, Seramikler", *Bilimve Teknik,* vol. 27, no. 317, pp. 6-15, 1994.

[21]   S. Parrott, "Engineering Ceramics for Aggressive Environments", *Met. Mater.,* vol. 6, no. 4, pp. 207-210, 1990.

[22]   F. Muilwijk, and J.P.P. Tholen, Ceramic Membranes, Applications and Properties.*Euro Ceramics.,* G. Ziegler, H. Hausner, Eds., vol. Vol. 3. Trans Tech Zürich Switzerland, 1993, pp. 3.596-3.599.

[23]   Ceramics in Military Applications., "Ceramic Fact Sheets", *The American Ceramic Society..* Web Site: http://www.acers.org/info/facts/facts.html

[24]   J.E. Sheenan, K.W. Buesking, and B.J. Sullivan, "Carbon-Carbon Composites", *Annu. Rev. Mater. Sci.,* vol. 24, pp. 19-44, 1994.
[http://dx.doi.org/10.1146/annurev.ms.24.080194.000315]

[25]   S. Nunomura, J. Nakayama, H. Abe, O. Kamigaito, and K. Matsusue, Mechanical Properties.*Advanced Technical Ceramics.,* S. Somiya, Ed., Academic Press San Diego: CA, 1989, pp. 223-258.
[http://dx.doi.org/10.1016/B978-0-12-654630-9.50016-4]

[26]   S. Nunomura, J. Nakayama, H. Abe, O. Kamigaito, and K. Matsusue, Mechanical Properties.*Advanced Technical Ceramics.,* S. Somiya, Ed., Academic Press San Diego: CA, 1989, pp. 223-258.
[http://dx.doi.org/10.1016/B978-0-12-654630-9.50016-4]

[27]   *The Morgan Crucible Company: A Leader in Ceramics Technology* Automotove Technology International: Sterling London, 1998.

CHAPTER 7

Advanced Ceramics for Effective Electromagnetic Interference Shields

U. Naresh[1], N. Suresh Kumar[2], K. Chandra Babu Naidu[3,*], A. Manohar[4] and **Khalid Mujasam Batoo[5]**

[1] *Department of Physics, BIT Institute of technology, Hindupuram-515502, A.P, India*

[2] *Department of Physics, JNTUA, Anantapuramu-515002, A.P, India*

[3] *Dept. of Physics, GITAM Deemed to be University, Bangalore-562163, Karnataka, India*

[4] *Department of Materials science and engineering, Korea University, 145 Anam- ro, Seongbuk-gu, Seoul 02841, Republic of Korea.*

[5] *King Abdullah Institute For Nanotechnology, King Saud University, P.O. Box 2455, Riyadh, 11451, Saudi Arabia*

Abstract: The electromagnetic (EM) interference is one of the major issues in EM wave applications. To prevent electronic devices from the EM wave interference, there is a need to find novel materials for shielding the EM wave interference. In this chapter, we discussed the EM wave interference and the effective EM interference parameters to find the strength of shielding. The chapter contains the mechanism and some derived materials reported by the various scientists. The materials are classified as ferrites, composites, and carbon-based materials. Herein, some other interesting materials like ferrites and conducting polymer composites were discussed to shield the EM interference.

Keywords: Dielectric loss, Electromagnetic Interference, Ferrites, Polymers, Shielding.

7.1. INTRODUCTION

Generally, the interference can be defined as the superposition of two or more waves thereby the modification of intensity. The modification contains minimum and maximum positions. If the waves are electromagnetic in nature, the applications of EM waves will be generally varied. The high-speed digital circuits and analog circuits of electronic systems are accounted for as major problems of electromagnetic interference (EMI). In recent years, the large-scale usage of elect-

* **Corresponding author Dr. K. Chandra Babu Naidu:** GITAM Deemed To Be University-Bangalore Campus, Bangalore-562163, Karnataka, India; E-mail: chandrababu954@gmail.com

K. Chandra Babu Naidu & N. Suresh Kumar (Eds.)

ronic systems in different fields, electromagnetic microwave interference is emerging as a new class of environmental pollution. It is creating significant damage to electronic devices and biological things like the human body and all the environmental species. In order to reduce the electromagnetic wave interference, to keep the human body and clean environment away from EMI radiation, considerable research has been focused on the improvement of composites such as ferrites, polymers and inorganic materials which are capable to absorb the microwave radiation thereby electromagnetic shielding capacity [1]. In last decade, a number of ideas were developed to reduce this type of EM pollution. These ideas include shielding, arranging the components of electronics as keeping away from radiation. The usage of embedded capacitance and components drop more data throughout the system. Among these methods, the preparation of shielding materials is the most useful method [2]. The consumption of the materials in regular life and industrial purposes depends on the constant supply of those materials.

Day by day, the consumption rate is growing exponentially owing to increase of population. Hence, for the sustainable future, it is important to find new ways to use those materials effectively. At present, the global attention is to overcome the challenges for sustainable advancement in materials. For this purpose, the industries forced to conduct self-assessment to identify the strategies and opportunities to achieve this goal. In addition, the strategy for the environmental sustainability is a long-term view *i.e.*, adjustment of the life style which meets the present requirements deprived of compromising the requirements of future generations. Due to continuous increasing demands for the ecofriendly products, particularly on dealing with products end of life phase, product designers and manufacturers are essential to consider the future disposal of their products. Further, well functioned recycling techniques were existed for the traditional materials like aluminium and steel. But these are not suitable for more extensively used structures of composites [3]. This is attributed to a fact that the composite material consists of several types of materials as a mixture on macro level. They may not be considered as homogeneous as the steel materials. These situations obscure the opportunities to form an efficient approach for waste management. Even though, several methods are existed, they are not commercially available. Landfill and incineration are the present disposal techniques of composite materials. Besides, many investigations are pointed out recycling of composite materials as the best alternative considering environmental effects. Exclusively, recycling of polymer composites is significant method to obtain complete information about the elements of these materials.

As discussed above electromagnetic microwave interference is being emerging as a new class of environmental pollution creating a significant damage to electronic applications and the biological things like human body and all of environmental species [4]. In order to target the electromagnetic interference, to keep safe from damages of equipment, keep the human body and clean environment far away from EMI radiation, a considerable research is being focused on the improvement of composites and materials. They are ferrites, polymers and inorganic materials which have capable to the microwave absorption and electromagnetic shielding capacity. Apart from these electromagnetic parameters, the high abundance of physical and morphological parameters such as low density, porous structure and multiple polarization centers can enhance the microwave absorption property. The high magnitude of the above said parameters can give rise to the multiple reflection and scattering of microwaves thereby matching the impedance [1, 3]. From the vast literature survey, it was evidenced that the ceramic materials belonging to the spinel ferrites, magnetic spinels, hexaferrites, magnetic perovskites, ceramic composites and ceramic polymer composites showed the high microwave absorption efficiency [2]. In view of this, the nickel zinc ferrites, nickel zinc cobalt ferrites, nickel cobalt zinc lanthanum ferrites, nickel zinc lanthanum ferrites, manganese zinc ferrites, barium zinc cobalt U-hexaferrites, barium hexaferrites, aluminium titanate and lanthanum iron oxide offered considerable microwave absorption behavior at giga hertz frequency range [2]. Once the microwave absorber materials are exposed to the EM radiation, EM radiation enters the interior of material. Later on, the electromagnetic radiation will be absorbed and transformed into dissipated energy. Therefore, the electronic object will be prevented from EMI. A photograph of microwave absorber is provided in (Fig. **7.1**) [5].

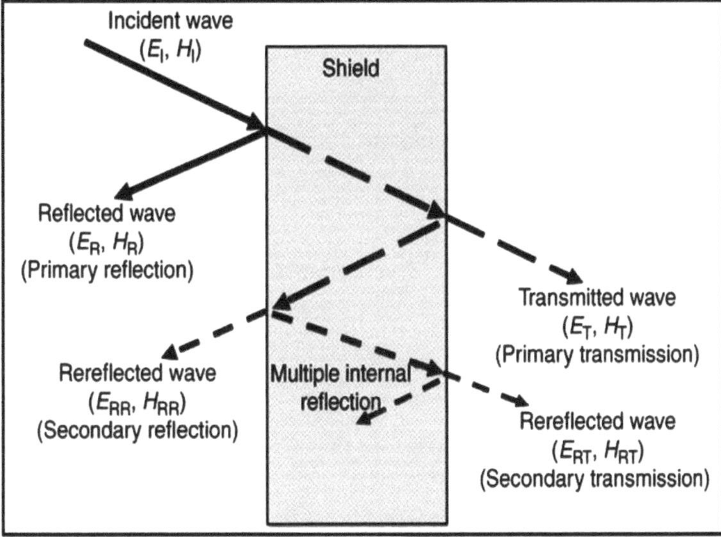

Fig. (7.1). Shielding mechanism of electromagnetic waves at the boundary.

7.2. EMI SHIELDING MECHANISMS

Let us consider a plane sheet which separates the source and receiver as the simple example for EMI shielding. The shielding is a method to decrease the interference of electromagnetic waves in the electronic equipments using a well designed material shield which is synthesized from the ferrites and polymers or their composites. The value of the shield is restrained as a function of effectiveness of the shielding (SE). This is logarithmic measure of the field strength in the absence and presence of the plane sheet for both electric and magnetic fields. Scientifically, shielding effectiveness of the shield is stated as [5 - 7]:

$$SE_T = SE_R + SE_A + SE_M \qquad (7.1)$$

Analysis of total shielding effectiveness SE_T discloses the elementary operation through which the substance affords the following shield-reflection (SE_R), shield-absorption (SE_A) and multiple internal reflections (SE_M) in sublattices in materials as shown in Fig. (**7.1**) [5].

Experimentally, SE_T is estimated from S-parameters (S_{11} or S_{22} and S_{12} or S_{21}) and they are obtained using the device of two ports VNA (vector network-analyzer) in decibels dB. These scattering parameters are connected to transmittance (T) and reflectance (R) of the signal from the shield as in case $SE_A > 10$ dB, SE_M is almost negligible [34] and overall SE_T might be stated as

$$SE_T = SE_R + SE_A \qquad (7.2)$$

Furthermore, the intensity of the signal $(1-R)$ enters the shield reflection which is utilized for standardization of absorption (A) to generate actual absorbance as

$$A_{eff} = (1-R-T)/(1-R) \qquad (7.3)$$

Hence, practical losses of absorption and reflection can be quantified as [8]:

$$S_{ER} = 10\log10\,(1 - R) \qquad (7.4)$$

$$S_{EA} = 10\log10(1- A_{eff})$$

(7.5)

It is clear from the theoretical investigations that the SE_T is a measure of the distance from the incident wave source. Mentioning to (Fig. **7.1**) [5], the electromagnetic region may be split into three portions depending on relative distance (d) of radiated signal wavelength. In the electromagnetism, the areas of the distance $d>\lambda/2\pi$ and $d<\lambda/2\pi$ are the far field and near field. The design of the EMI shielding is carried out for the specific shielding application. In addition, the shielding effectiveness of planar shield is governed by different mechanisms which include contact between the incident wave with air interface, shield interface and the shield medium. Besides, the portion of the loss of reflection is defined as the wave reflection from shield interface due to mismatch of the impedance between surface impedance of the shield and incident wave. The loss of reflection for a plane wave is stated as [8]:

$$SE_R=-10\log_{10}\sigma /16\omega\varepsilon\mu_r$$

(7.6)

The loss of absorption of a wave is referred as the propagation of the wave in the thickness of the shielding medium (t) gets weakened by transforming the radiative energy into thermal energy owing to induced current inside the medium. The expression $E_t = E_i e^{-t/\delta}$ represents the electric field and $H_t = H_i e^{-t/\delta}$ represents the magnetic field within the shield, where δ is the depth of the skin described as the required distance of the wave to attenuate $1/e$ or 37% of its original value. Precisely, skin depth is written as:

$$\delta = \frac{1}{\sqrt{\pi\mu f\sigma}}$$

(7.7)

Here 't' indicates the thickness of the shield. The above equation reveals that for good absorption of the signal of frequency (f), the shielding material should have large conductivity and large permeability.

Multiple reflection loss is a percentage of the wave reflected from the second boundary. Besides, it yields to second boundary to be reflected again as illustrated in **Fig. 7.1** [5]. It is another mechanism that occurs in within the shielding medium. These reflections cause an extra loss in the medium that increases the shielding effectiveness. Nevertheless, the substantial attenuation as the preliminary wave propagates through shield consisting of thickness of several skin depths reduces the effect of the multiple reflections. Scientifically, loss due to multiple reflections can be stated as [8]:

$$SE_M = 20\log_{10}(1 - e^{-2t/\delta}) \qquad\qquad (7.8)$$

Therefore, from the theoretical analysis of SE_T, it was revealed that the frequency dependent SE_T parameters are the thickness, permittivity, conductivity and permeability of the shielding medium. Which can be measured form the scattering parameters using various methods [9].

7.3. EMI SHIELDING MATERIAL

To prevent the electronic devices from the electromagnetic interference, the researchers interested very much in the direction of finding the effective material synthesis whose total shielding efficiency is very high. The other related properties are acquired to minimize increase of EM interference. Generally, the ferrite material is one of the very good materials for the EM shielding purpose. In addition to these materials, the researchers focused on their composites for the compatibility and flexibility. There are the novel materials such as soft and hard ferrites. The polymer composites and carbon composites with the ferrites are other novel materials for the application of electromagnetic shielding.

7.4. FERRITE BASED COMPOSITES

The microwave absorbing materials can be fabricated by loading different types of polymeric binders in various dielectric and magnetic materials which are in powder form. Further, Ni-Zn based spinel ferrites can be utilized for the high frequency applications. Especially, in case of antennas, transformer cores, radio frequency [5, 6] and radar absorbing materials [8]. Good magnetic characteristics, fine particle size and high homogeneity of the materials can be achieved by synthesizing these materials with wet chemical techniques [7–9]. In addition, it is proved that the addition of CuO to the NiZn-ferrite reduces the sintering temperature and changes the magnetic properties [10]. Besides, the doping of Cu in Ni-site also affects the resistivity and bulk density of the ferrite. Furthermore, with the help of citrate precursor method Nicolson *et al.* [10], synthesized NiZn and NiCuZn ferrite nanoparticles and studied the influence of Cu on the spinel phase of NiZn ferrite. They reported that electromagnetic and absorption properties of prepared ferrites in range of 1 MHz–1.8 GHz and 7.5–13.5 GHz at low temperatures.

Aphesteguy *et al.* [11], prepared the Cu substituted NiZn-ferrite and they observed the improvement of permeability, magnetic loss and saturation magnetization at 1.3 GHz as 0.01 in NiCuZn ferrite. On the other hand, it is very small (0.006) for NiZn ferrite. Also, the NiCuZn-ferrites exhibit the excellent absorption properties between the frequencies 1 MHz and 1.8 GHz, whereas NiZn-ferrite showed the highest loss of reflection about 1.8 dB at 0.7 GHz. The absorption properties of NiZn ferrite over 8–12 GHz frequency range are suitable for random access memories. Further, the low value of coercivity directed to mobilization of domain wall that absorbs the magnetic energy and converts it into heat. Majeeda *et al.* [12], synthesized the $Sr_2Ni_2T_xFe_{28-x}O_{46}$ ($T=Cr^{3+}$, Bi^{3+}, Al^{3+}, In^{3+}; x=0.25) hexaferrites by sol-gel technique. In between the frequencies 0.001 GHz and 1 GHz, they observed the increment in dielectric constant along with dielectric loss for all samples. Particularly, at high frequencies the Al^{3+} and In^{3+} substituted materials exhibit high dielectric constant. The resistance at grain and grain boundaries within the material may be the reason for achieving high values of dielectric constant and loss. These materials with high impedance and dielectric constant with moderate dielectric loss can serve as high frequency microwave absorbers. Furthermore, the analysis of magnetic properties revealed that due to high values of retentivity and coercive force, the Al^{3+} and In^{3+} substituted samples are potential candidates for the microwave absorbing applications, recording/storage applications and formation of multilayer chip inductors.

Using the sol-gel auto-combustion method, Dhruva *et al.* [13], synthesized the Ga-substituted Z-type $Sr_3Co_{2-x}Ga_xFe_{24}O_{41}$ (x = 0.0, 0.4, 0.8, 1.2, 1.6 and 2.0) hexaferrites. The presence of Z, W, Y and M phases was confirmed by XRD studies. The prepared ferrite materials exhibit typical soft magnetic properties. Further, the saturation magnetization lies in the range of 64-76 Am^2kg^{-1}. The replacement of the Ga in the prepared hexaferrites leads to reduction in the values of saturation magnetization. The reduction in saturation magnetization with substitution of Ga is in good agreement with analysis of Mössbauer spectroscopy. The hyperfine magnetic field of all samples was noticed to be diminished and also the increment on the relative area of the doublet. In addition, the analysis of Mössbauer spectroscopy revealed that for the compositions below $x \leq 0.4$, the Fe ions exhibit high spin rate of (Fe^{3+}), whereas almost 1.5% of the Fe ions are transformed to Fe^{2+} spin state for the compositions above $x \geq 0.8$. Therefore, the Ga^{3+} ions replaced Fe^{3+} ions instead of remaining ions. Besides, at low frequencies, the prepared samples exhibit normal dielectric response which is described through Maxwell-Wagner's model. Further, in microwave region the prepared samples exhibit the relative permittivity about 8-14. In particular, at the frequency 11 GHz, the x=0.0 composition exhibits ferromagnetic resonance and at the frequency of 12.15 GHz, the x=2.0 composition exhibits the dielectric resonance. Hence, at the specific microwave frequencies, these materials can serve as EMI shielding and microwave absorbers. In addition, above the X-band frequencies, the reflectivity of the prepared samples having thickness 3 mm differs from -8 dB to 2.3 dB. This evidenced that the x = 2.0 (at 12.15 GHz) shows both dielectric resonance and FMR by coupling two properties, there is a chance to develop metamaterials.

7.5. CARBON BASED COMPOSITES

In comparison with the conventional carbon fillers, the carbon nanotubes and carbon nanofibers exhibit remarkable electrical, mechanical and morphological properties like larger conductivity, larger filed strength, high aspect ratio and small diameters. Further, the utilization of these nanostructured carbons as a filler in polymer composites permits the structures with lower filler loadings to afford anticipated electrical and electromagnetic shielding properties [14]. The quantity of carbon nanotubes in the composite is a significant issue because carbon nanotubes are expensive. The high amount of CNTs is required for compressed molding technique. The polymer matrix can be minimized after adding the percentage of CNTs. These composites are versatile and have wide range of practical applications. Screen printing method is also suitable for preparing EMI materials. This is a technique which is mainly suitable for flat surfaces compared to compressing molding technique. The screen-printing technique is inexpensive, simple, controlled thickness and homogeneity of the films. Different carbon materials (Figure. 2) in reference [14] illustrates the SEM pictures of CNTs and graphite with slight aspect ratio and CNTs with high aspect ratio which are prepared by screen printed technique for EMI shielding applications.

In general, the carbon-based (CB) materials show various morphologies. The CB is to be small and homogeneous particle while graphite has sheet like morphology. In addition, the shape of both CNTs is tubular with distinct length and diameter. Owing to small length and larger diameters of CNTs, the aspect ratio (L/D) is small, whereas the value of L/D is 1000 in which the diameters of the CNTs are very small. Besides, all the prepared mono-printed layered and multi-printed layered carbon films are very smooth and homogeneous. They reported that the qualitative measurement of adhesion printed carbon nanotubes to the material can be done through shaking and dropping the samples followed by bending. Due to the flexible plastic substrate, the carbon films adhered to the substrate even after bending process. Further, they observed that in printed materials, there will be no loss due to coatings and qualitatively indicating a decent bonding strength among the printed carbon films and the substrate which was confirmed by simple peel test utilizing adhesive tape. The carbon films are still adhered to the substrate even after removing the adhesive tape with negligible disturbance on the surface of the film. The shielding effectiveness of various carbon materials is illustrated in Figure. 3 of reference [14] over 15 MHz to 1000

MHz. Furthermore, the printed CB and CNT sheet of thickness 100 mm exhibits considerable shielding effectiveness greater than 10 dB. Curiously, the two types of CNTs afford larger EMI shielding than CB and graphite while all printed films comprise the same concentrations of 15 wt % of carbon fillers. However, it is well known that in case of conductive composite, the effectiveness of EMI shielding is strongly allied with their dc-conductivity. Shielding effectiveness is defined as SE (dB): $10\log(P_t/P_o)$, where P_t is transmitted power and P_o is the incident power. The SE is increased with respect to increment in electrical conductivity of shielding material. This depends on the EMI shielding theory. Therefore, shielding effectiveness is exhibited by the CNTs which are having large aspect ratios. The significant dispersion of the CNTs is observed in the printed MWCNTs *i.e.*, both in short and the long MWCNT films. These MWCNTs are consistently spread with few tangles and aggregates in the printed film. It is clear from the SEM pictures that the tube like CNT structure can establish a conductive network for transmission of electrons. The distance between the adjacent conducting clusters in CNTs are small compared to the other structures of CNTs such as sheet or particle forms. This is happened due to separation of the neighboring clusters through the organic insulating materials. Besides, the conduction mechanism of the two electrons might be elucidated through quantum-tunneling, in which the height of the barrier reduces with increase in temperature. Hence, the conduction of the printed composites comprising of conductive fillers is extremely affected by separation of adjacent conducting clusters or conductive fillers [15]. In addition, in comparison with MWCNTs, the graphite and CB powders have comparatively large sizes and low aspect ratios, which are inadequate to establish conductive networks. Further, the value of SE for short MWCNT-films is three times less than that of long MWCNT-films. Besides, the aspect ratio of shorter CNT is 10 times less than the longer CNT. In addition, group of researchers studied the electromagnetic interference shielding effectiveness and electrical properties of the multi-walled carbon nanotubes/polycarbonate (MWCNT/PC) composites.

By using screen printing technique, Wang *et al.* [16], fabricated a carbon-based film for the applications in EMI-shielding. The outcomes exhibit that due to the improved electron transmission, the CNTs can act as effective EMI shielding absorbers in broad band frequency range in comparison with graphite and carbon black. Further, the shielding performance of thin printed CNT film (150 mm thin and 15 wt % of CNT) is almost analogous to the thick epoxy composite CNT film (1.5 mm thick and 15 wt % of CNT). From this, it is clear that screen printing technique might be a promising technique for fabricating commercial EMI shielding thin films. Furthermore, Gupta *et al.* [17], fabricated effective light weight PTT/MWCNT (Poly(tri-methylene terephthalate)/multi-walled carbon nanotube) composites with varying amounts of MWCNTs to serve as effective

EMI shielding material in Ku-band frequencies *i.e.*, 12.4-18 GHz. At 10% (w/w) MWCNT loading, the PTT/MWCNT composite shows the shielding effectiveness (SE_T) about 36–42 dB. The shielding mechanism was investigated through resolving the total shielding effectiveness into reflection loss (SE_R) and absorption (SE_A). Besides, the PTT/MWCNT composite exhibits dominating shielding properties. Consequently, it might be utilized in RADAR and microwave absorbing material. In addition, the loading of MWCNTs affects the dielectric properties and electrical conductivity of PTT. By using diluting master batch (15 wt % MWCNT), Mohmmad *et al.* [18], prepared the MWCNT composites followed by Haake mixer and then injection-molded into a dog-bone mold. Under different working conditions, several MWCNT alignments have been created. Besides, the magnitude of electrical resistivity is measured at three distinct areas in the directions parallel and perpendicular to the flow direction. From the results, it is clear that the resistivity and filtration is maximum at high dense regions in parallel and perpendicular directions with respect to direction of the flow. After filtration, at larger alignments, the field emission mechanism is more significant which was observed by applying Obm's law. The SEM, TEM and Raman spectroscopy analysis confirms the presence of larger MWCNT orientations various fields with larger resistivities. Furthermore, EMI SE measurements were carried out for the prepared samples at different concentrations and thicknesse by compression-molded technique. The outcomes revealed that with respect to increase in MWCNT loading and thickness, the EMI shielding effectiveness (*i.e.*, both by reflection and absorption) of the material is increased.

7.6. CONDUCTING POLYMER BASED COMPOSITES

In general, the ferromagnetic conducting polymer composites exhibit both electrical and magnetic properties that can be achieved by combining the magnetic nanoparticles with conducting polymers. This unique property of the polymer composite permits using the fabrication of electromagnetic shielding materials in case of the electric field and magnetic fields developed at right angles to each other. The electromagnetic waves produced from both sources (electric and magnetic) can be shielded efficiently by using conducting ferromagnetic type materials. As we know that in EMI shielding, the primary mechanism is reflection. On the shield, there exists mobile charge carriers *i.e.*, electrons and holes for the reflection of the electromagnetic radiation by the shield. These charge carriers can interact with the electric and magnetic fields of the incident radiation. Therefore, the shield tends to be electrically-conducting, even though high conductivity is not essential. Furthermore, the absorption is the secondary mechanism in the EMI shielding. Also, to interact with the electric and magnetic fields of the incident radiation, the shield must have the electric and magnetic dipoles in order to observe significant absorption. Hence, the single system containing both electric and magnetic components can be utilized as an EMI shielding material.

Based on the major physical parameters such as the pulsation ($\omega=2\pi f$), electrical-permittivity (ε_r) and the magnetic-permeability (μ_r), the material can be categorized as prominent microwave absorbing material or not. From the recent reports, it is evident that the non-conducting polymer composites filled with conducting materials such as CNTs, nickel *etc.*, can possess promising shielding performance [19]. Nevertheless, to attain the saturation level which leads to microwave losses, the high filler contents are required *i.e.*, for fiber it is about 10-15% and for spheres, it is about 30%. In general, multilayer structures are necessary for the design of large bandwidth structures for which the frequency dependence of imaginary part of permittivity must be controlled [20]. However, the usage of dielectric materials like carbon-black, graphite fibers, metallic powders *etc.*, which are attained by conductive filler dispersion is limited. Moreover, particle aggregation phenomena and filler content effectively influence the performance of the material [21]. Besides, the complex permittivity is generally depending on the processing conditions and the texture of the percolations. The threshold percolation varies with respect to the shape of the particle (fibers, plates & sphere). At the time of processing, the filler accumulates with nano-metric particles like carbon-black or micron sized fillers like carbon-sphere lead to difficulty in material reproducibility. To overcome these drawbacks, various techniques have been proposed. In this connection, the conducting polymer composites gained much attention and exhibit outstanding properties owing to the chemical nature of thee macromolecular chains in which electronic-conduction occurs at broad range. Also, as we compared to the traditional dielectric materials, the conducting polymer composites have special features which make them more prominent compared with others [22]. The chemistry of these conducting polymers offers numerous methods for synthesis which have good reproducible properties. Also, the insertion of conducting polymers into different materials such as insulating polymer matrix, honeycomb structure and reinforcing fabrics induced complex structures.

In addition, by using mechanical blending method, Tchouank *et al.*, [23] synthesized Co_2Y/BCTO/PANI composites. They reported that the prepared composites exhibit pure phases of Co_2Y barium hexaferrites and bismuth copper titanate (BCTO) which was confirmed by XRD and Raman analysis. Further, in the spectra of Co_2Y/BCTO/PANI composites, the resonance ensures the entire frequency region *i.e.*, owing to interfacial polarization, the dielectric losses dominate in low frequency region, whereas at high frequency region magnetic losses dominate owing to eddy current losses. Moreover, the measured values of

SE_R are larger than that of measured values SE_A. This is an indication that most of the incident radiation undergoes reflection rather than absorption [24]. The reported values showed that the effectiveness of the shielding for the prepared composites is 31.97 dB at 10.30 GHz with thickness of 2 mm. As compared to magnetic loss, the Co_2Y/BCTO/PANI composite exhibits larger dielectric losses. Besides, the total shielding effectiveness value is accredited to excellent impedance matching among the components of Co_2Y/BCTO/PANI composite. In order to explore the influence of magnetic fillers on shielding properties of conductive epoxy nanocomposites, Varun *et al.* [25], prepared these epoxy-based hybrid nanocomposites consisting of different ratios of Fe_3O_4 and/or CB *via* melt blending technique. From the EMI shielding analysis, in the X-band frequency region the prepared composites with 6 wt % of CB and 5 wt % of Fe_3O_4 unveiled an EMI SE of 8.5 dB. Also, the composites consisting of 6 wt % of CB was noticed to be increased with respect to Fe_3O_4. The SEM pictures attained on the freeze-fractured cross-section of the prepared materials disclose appropriate filler distribution in the prepared epoxy composite. Besides, the Young's modulus of the epoxy-composites was noticed to be increased with increase of composition. Further, the prepared composites with 6 wt % of CB have the Young's modulus about 2.16 GPa and 6 wt % of CB/3 wt % of Fe_3O_4 have the Young's modulus about 2.73 GPa. From these results, it is clear that the addition of magnetic fillers enhances the shielding properties of prepared conducting polymer composites. Sambyal *et al.* [26], have synthesized conducting-polymer composite entrenched with hybrid material (BST/RGO/Fe_3O_4). They reported that in the prepared conducting matrix, the filler materials are homogeneously distributed which was confirmed by SEM analysis. Further, the prepared composite exhibits high value of shielding effectiveness about 48 dB in X-band frequency range (8.2-12.4 GHz). This is observed due to the mixture of dielectric and magnetic filler materials in conducting polymer composite. Hence, the composites with large absorption properties are the promising candidates for next generation EMI shielding materials.

CONCLUSIONS

The research on the electromagnetic shielding materials was increased from the last few decades. The materials commonly used are ferrites and their composites to shield the EM wave. The total effective shielding depends upon the shielding reflectance, absorbance, and multiple reflections between the layers of the shielding materials. In addition to this, the material permeability and dielectric loss are also the deciding properties of the shielding effectiveness of the materials. The layers of the materials manufactured with the help of ferrite based materials like hexaferrites and their based materials, carbon-based materials like carbon nanotubes and their related substances. Moreover, the polymer-based materials, conducting polymer composites with ferrites and their mechanism were discussed.

CONSENT FOR PUBLICATION

Not applicable.

CONFLICT OF INTEREST

The authors confirm that this chapter content has no conflict of interest.

ACKNOWLEDGEMENTS

The authors express thankfulness to Dr. P. Sreeramulu, Assistant Professor (English), GITAM, Bangalore for providing English language editing services to this manuscript.

REFERENCES

[1]   K.C.B. Naidu, and W. Madhuri, "Microwave Processed Bulk and Nano NiMg Ferrites: A Comparative Study on X-band Electromagnetic Interference Shielding Properties", *Mater. Chem. Phys.,* vol. 187, pp. 164-176, 2017.
       [http://dx.doi.org/10.1016/j.matchemphys.2016.11.062]

[2]   X.C. Tong, *Advanced Materials and Design for Electromagnetic Interference Shielding.* CRC Press, 2009.

[3]   J.S. Ghodake, C.K. Rahul, T.J. Shinde, P.K. Maskar, and S.S. Suryavanshi, "Magnetic and microwave absorbing properties of Co2+ substituted nickel–zinc ferrites with the emphasis on initial permeability studies", *J. Magn. Magn. Mater.,* vol. 401, pp. 938-942, 2016.
       [http://dx.doi.org/10.1016/j.jmmm.2015.11.009]

[4]   X.T. Vuong, "Military X-band very small aperture terminals (VSATs)—to spread or not to spread, in: Mili-tary Communications Conference, MILCOM'96", *Conference Proceedings,* 1996pp. 6-10

[5]   Y. Huang, N. Li, Y. Ma, F. Du, F. Li, X. He, X. Lin, H. Gao, and Y. Chen, "The influence of single-walled carbon nanotube structure on the electromagnetic interference shielding efficiency of its epoxy composites", *Carbon,* vol. 45, pp. 1614-1621, 2007.
       [http://dx.doi.org/10.1016/j.carbon.2007.04.016]

[6]   C.R. Paul, *Electromagnetics for Engineers.* Wiley: Hoboken, New Jersey, 2004.

[7]   P. Saini, V. Choudhary, K.N. Sood, and S.K. Dhawan, "Electromagnetic interference shielding behavior of polyaniline/graphite composites prepared by *in situ* emulsion pathway", *J. Appl. Polym. Sci.,* vol. 113, no. 5, pp. 3146-3155, 2009.
       [http://dx.doi.org/10.1002/app.30183]

[8]   R. Che, L.M. Peng, X.F. Duan, Q. Chen, and X.L. Liang, "Microwave absorption enhancement and complex permittivity and permeability of Fe encapsulated within carbon nanotubes", *Adv. Mater.,* vol. 16, no. 5, pp. 401-405, 2004.
       [http://dx.doi.org/10.1002/adma.200306460]

[9]   V. Choudhary, S.K. Dhawan, and S. Parveen, Polymer based nanocomposites for electromagnetic interference (EMI) shielding.*EM Shielding Theory and Development of.,* M. Jaroszewski, J. Ziaja, Eds., New Materials, Research Signpost: Kerala, India, 2012, pp. 67-100.

[10]　A.M. Nicolson, and G.F. Ross, "Measurement of the intrinsic properties of materials by time-domain techniques", *IEEE Trans. Instrum. Meas.,* vol. 19, no. 4, pp. 377-382, 1970.
[http://dx.doi.org/10.1109/TIM.1970.4313932]

[11]　J.C. Aphesteguy, A. Damiani, D. DiGiovanni, and S.E. Jacobo, "Microwave-absorbing characteristics of epoxy resin composites containing nanoparticles of NiZn- and NiCuZn-ferrites", *Physica B,* vol. 404, pp. 2713-2716, 2009.
[http://dx.doi.org/10.1016/j.physb.2009.06.065]

[12]　A. Majeed, M.A. Khan, L.M. Yousuf, R. Ahmad, and I. Ahmad, "Structural, microwave permittivity, and complex impedance studies of cation (Cr, Bi, Al, In) substituted SrNi-X hexagonal nano-sized ferrites", *J. Alloys Compd.,* vol. 46, pp. 1907-1915, 2020.

[13]　P.N. Dhruv, S.S. Meena, R.C. Pullar, F.E. Carvalho, R.B. Jotania, P. Bhatt, and C.L. Prajapat, "Joã. Paulo. Barros Machado, T.V. Chandrasekhar Rao, Basak C.B., Investigation of structural, magnetic and dielectric properties of gallium substituted Z-type Sr3Co2-xGaxFe24O41 hexaferrites for microwave absorbers", *J. Alloys Compd.,* vol. 822, 2020.153470
[http://dx.doi.org/10.1016/j.jallcom.2019.153470]

[14]　D.R.J. White, *EMI/EMC handbook series.* Don White Consultants Inc.: New York, 1971, pp. 4-0.

[15]　C.S. Zhang, Q.Q. Ni, S.Y. Fu, and K. Kurashi, "Electromagnetic interference shielding effect of nano-composites with carbon nanotubes and shape memory polymer", *Combust. Sci. Technol.,* vol. 67, pp. 2973-2980, 2007.
[http://dx.doi.org/10.1016/j.compscitech.2007.05.011]

[16]　N. Joseph, C. Janardhanan, and M.T. Sebastian, "Electromagnetic interference shielding properties of butyl rubber-single walled carbon nanotube composites", *Compos. Sci. Technol.,* vol. 101, pp. 139-144, 2014.
[http://dx.doi.org/10.1016/j.compscitech.2014.07.002]

[17]　L-L. Wang, and B-K. Ta, "Kye-Yak See, Zhuo Sun, Lin-Kin Tan, Darren Lu Electromagnetic interference shielding effectiveness of carbon-based materials prepared by screen printing", *Carbon,* vol. 47, pp. 1905-1910, 2009.
[http://dx.doi.org/10.1016/j.carbon.2009.03.033]

[18]　A. Gupta, and V. Choudhary, "Electromagnetic interference shielding behavior of poly (trimethylene terephthalate)/multi-walled carbon nanotube composites", *Compos. Sci. Technol.,* vol. 71, pp. 1563-1568, 2011.
[http://dx.doi.org/10.1016/j.compscitech.2011.06.014]

[19]　M. Arjmand, G.A. Gelve, and S. Park, "Uttandaraman, Sundararaj Electrical and electromagnetic interference shielding properties of flow-induced oriented carbon nanotubes in polycarbonate", *Carbon,* vol. 49, pp. 3430-3440, 2011.
[http://dx.doi.org/10.1016/j.carbon.2011.04.039]

[20]　H. Zhang, G. Zeng, Y. Ge, T. Chen, and L. Hu, "Electromagnetic characteristic and microwave absorption properties of carbon nanotubes/epoxy composites in the frequency range from 2 to 6 GHz", *J. Appl. Phys.,* vol. 105, 2009.054314
[http://dx.doi.org/10.1063/1.3086630]

[21]　B. Gao, L. Qiao, J. Wang, Q. Liu, F. Li, J. Feng, and D. Xue, "Controllable synthesis, characterization and microwave absorption properties of magnetic Ni1−xCoxP alloy nanoparticles attached on carbon nanotubes", *J. Phys. D Appl. Phys.,* vol. 41, 2008.125303
[http://dx.doi.org/10.1088/0022-3727/41/12/125303]

[22]　M.H.A. Saleha, and U. Sundararaj, "Electrically conductive carbon nanofiber/polyethylene composite: effect of melt mixing conditions", *Polym. Adv. Technol.,* vol. 22, pp. 246-253, 2011.
[http://dx.doi.org/10.1002/pat.1526]

[23]　J.G. Park, J. Louis, Q. Cheng, J. Bao, J. Smithyman, R. Liang, B. Wang, C. Zhang, J.S. Brooks, L.

Kramer, P. Fanchasis, and D. Dorough, "Electromagnetic interference shielding properties of carbon nanotube buckypaper composites", *Nanotechnology,* vol. 20, no. 41, 2009.415702
[http://dx.doi.org/10.1088/0957-4484/20/41/415702] [PMID: 19755727]

[24]   J. Tchouank Tekou Carol, "Mohammed, D. Basandrai, S. Kumar Godara, G.R. Bhadu, S. Mishra, N. Aggarwal, S.B. Narang, A.K. Srivastava, X-band Shielding of Electromagnetic Interference (EMI) by Co2Y Barium Hexaferrite, Bismuth Copper Titanate (BCTO), and Polyaniline (PANI) composite", *J. Magn. Magn. Mater.,* vol. 501, 2020.166433
[http://dx.doi.org/10.1016/j.jmmm.2020.166433]

[25]   V. Uma Varun, B. Rajesh Kumar, and C. Etika Krishna, "Hybrid polymer nanocomposites as EMI shielding materials in the X-Band", *Materials Today: Proceedings,* 2019pp. 300-305

[26]   P. Sambyal, S.K. Dhawan, P. Gairola, S.S. Chauhan, and S.P. Gairola, "Synergistic effect of polypyrrole/BST/RGO/Fe3O4 composite for enhanced microwave absorption and EMI shielding in X-Band", *Curr. Appl. Phys.,* vol. 18, no. 5, pp. 611-618, 2018.
[http://dx.doi.org/10.1016/j.cap.2018.03.001]

CHAPTER 8

Advanced Ceramics for Ferroelectric Devices

Prasun Banerjee[1,*], **Adolfo Franco**[2], **K. Chandra Babu Naidu**[1], **Thiago E.P. Alves**[3] and **R.J.S. Lima**[4]

[1] *Department of Physics, Gandhi Institute of Technology and Management (GITAM) University, Bangalore-561203, India*

[2] *Instituto de Física, Universidade Federal de Goiás, Goiânia, Brazil*

[3] *Instituto Federal de Educacao Ciencias e Tecnologia de Goias, Anapolis, Goias, Brazil*

[4] *Unidade Academica de Fisica, Universidade Federal de Campina Grande, Campina Grande, Paraiba, Brazil*

Abstract: The presence of intrinsic polarizations into the ferroelectric materials helps them to get polarized easily on the application of the electric field. The presence of the retentivity and the coercivity formed a hysteresis loop pattern when the field is applied to the materials. But some artifacts make the nature of such a hysteresis loop distorted for the ferroelectric materials. In general, the non-centrosymmetric nature of the crystal structure responsible for the ferroelectric nature of the materials. That is why the presence of the intrinsic polarization in the electrets does not make it a ferroelectric class of materials. Actually, the ferroelectrics come under the dielectric class of materials. The present-day application of the ferroelectric materials ranges from the ferroelectric memory devices, electro caloric devices, magneto electric devices, DRAM capacitors, *etc.* The flexible properties along with the high writing speed and cyclability make $Bi_{3.25}La_{0.75}Ti_3O_{12}$ as one of the most promising materials for FERAM applications. The large ECE values of the Barium hafnium titanate makes it very useful for the electro caloric cooling of the microelectronics devices. The presence of both spin glass state as well as the ferroelectricity make advanced ceramics like $La_3Ni_2NbO_9$ couple both magnetic and electric field within the same material for the fabrication of the magneto-electric devices. The presence of a morphotropic phase boundary in the advanced ceramics of hafnium oxide and zirconium oxide can result in a high dielectric constant for the DRAM applications.

Keywords: Ceramics, DRAM, Ferroelectric Materials, Magneto-Electric Devices, Magnetic loss.

* **Corresponding author Dr. Prasun Banerjee:** GITAM Deemed to Be University-Bangalore Campus, Bangalore-562163, Karnataka, India; E-mail: *pbanerje@gitam.edu*

8.1. INTRODUCTION

Ferroelectrics are the class of material with intrinsic polarization and this polarization can be tuned in accordance with the applied electric field [1 - 4]. When the external field is applied the polarization keeps on increasing due to the alignment of the electric dipoles along the direction of the applied fields [5]. When all the dipole alignment completes the polarization get saturated although the increase of the applied fields. But on the field reversal the material does not follow the same path and trace a loop known as the hysteresis loop [6 - 10]. But due to different artifacts the hysteresis loop can be different from their magnetic counterparts. For example, electrets normally build with melting the plastic or glass and thereafter polarized with applied external electric fields can also show a hysteresis loop but they can never be treated as ferroelectrics [11 - 13]. It is commonly known facts that although all the ferroelectrics dielectrics are insulators but we can treat all the insulators as dielectrics.

Actually, the dielectrics can be classified in different types for example ferroelectrics, pyroelectrics, piezoelectrics *etc*. Here the ferroelectrics are the core part of the all dielectrics. For instance, zinc oxide is a well-known pyroelectric material but it is not a ferroelectric material [14 - 16]. But on the other hand, barium titanate is a very popular ferroelectric materials hence by default it is also pyroelectric and piezoelectric in nature [17 - 20]. Hence, we cannot claim that there exists a center of symmetry for the ferroelectric material nor they are just simple glass material. Due to these features the ferroelectric hysteresis loop always comes with distorted shapes due to the losses along the artifacts.

In the earlier days soon after the discovery of ferroelectric materials in 1920 by Valasek people start believing that the hydrogen bond is the responsible factor. The reason for that was the water-soluble nature of the earlier days of ferroelectrics. But with the discovery of barium titanate in 1943 the real physics of non-centrosymmetric design of the ferroelectric start coming into the picture. Gradual shift of research towards ferroelectric thin film and there after successful integration to silicon chips a wide new area called ferroelectric memory devices comes into light in the year 2000 onwards. Not only ferroelectric memory devices [21, 22] we can see advanced ceramics applications in electro caloric devices [23, 24], magneto electric devices [25, 26], DRAM capacitors [27, 28], magnetic field sensors [29, 30], electron emitters [31, 32] and spatial light modulators [33, 34] *etc*. So, in this book chapter we will focus to understand the role of different ferroelectric ceramics for the device applications.

8.2. FERROELECTRIC MEMORY DEVICES

The uses of PZT and SBT are well known in the fabrication of nonvolatile ferroelectric random memory devices. But we will discuss here some advanced ferroelectric materials like $Bi_{3.25}La_{0.75}Ti_3O_{12}$ for such applications.

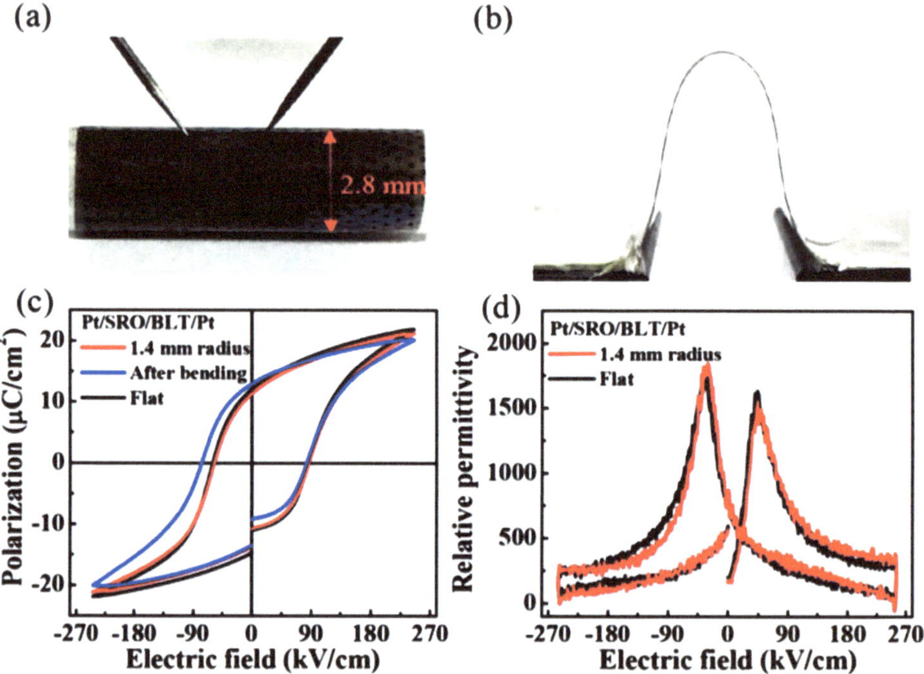

Fig. (8.1). (a & b) Sketches **(c)** P-E loops **(d)** ε_r-E curves of mica/Pt/SRO/BLT/Pt memories. Reprint with the permission from Ref [35]. Copyright 2018, ACS Publishing.

8.2.1. $Bi_{3.25}La_{0.75}Ti_3O_{12}$

Although, PZT films are very popular but they are, in general, fabricated in a silicon substrate that makes the films nonflexible. Hence for advanced flexible ferroelectric memory devices applications the used of PZT films are in general unsuitable. Hence the use of new advanced materials like MoS_2 is gradually coming into action. But there are certain drawbacks in these materials such as slow writing speed as well as lower cycles of use. Hence, very recently a new advanced ceramic material of $Bi_{3.25}La_{0.75}Ti_3O_{12}$ reported for flexible FERAM application purpose shown in Fig. (**8.1**) [35]. Here the mica has been used as substrate and Pt/SrRuO_3 used as electrodes. The thickness of the substrate can be reduced as thin as ten micron which enhanced the capabilities of the FERAM devices. Hence the fabricated device can withstand higher strain as well as the saturation polarization is also higher than that of the other materials. Not only that the reported cycle of use also more as well as the fatigue also less for wide spread industrial applications.

8.3. FERROELECTRIC ELECTROCALORIC DEVICES

Electrocaloric devices are the integral part for the microelectronics device cooling applications. Lead zirconium titanate is one of the most popular materials that can show electro caloric effect but the presence of lead in the material makes it toxic as well as the reported value of the conversion efficiency also lower. Hence, promising new advanced ceramics like barium hafnium titanate came into picture.

8.3.1. $Ba(Hf_xTi_{1-x})O_3$

Barium hafnium titanate prepared with the solid-state reaction technique showed large ECE values of 1.64 K for x=0.05 [36]. The comparative results of the recent progress in the field with different ceramics have been presented in the (Table **8.1**) Here, it can be observed that the electro caloric coefficient for the barium hafnium titanate *i.e.*, the value of the ratio of delta T and delta E is maximum among all the material reported so far.

Table 8.1. Comparison between different advanced ceramics materials for different electro caloric parameters. Reprint with the permission from Ref [36]. Copyright 2018, ACS Publishing.

Material	T_{Emax} (°C)	ΔT_m (K)	E (kV·cm-1)	$\Delta T/\Delta E$ (K·mm·kV-1)	ref
$Ba(Hf_{0.03}Ti_{0.97})O_3fp$	115	0.38	10	0.38	
$Ba(Hf_{0.03}Ti_{0.97})O_3fp$	119	1.34	50	0.27	
$Ba(Hf_{0.05}Ti_{0.95})O_3fp$	117	1.64	50	0.33	
$Ba(Hf_{0.05}Ti_{0.95})O_3sp$	76	1.21	50	0.24	[36] [37] [38]
$Ba(Hf_{0.11}Ti_{0.89})O_3fp$	92	1.41	50	0.28	[39]
$Ba_{0.94}Sm_{0.04}TiO_3$	76	0.92	30	0.30	
$BaTi_{0.98}Sn_{0.02}O_3$	85	0.16	9.84	0.16	
$Ba_{0.80}Sr_{0.20}TiO_3$	76	0.39	20	0.195	
$BaTi_{0.885}Sn_{0.105}O_3$	30	0.61	20	0.305	[40]
$Ba (Ti_{0.944}Y_{0.056})O_{2.972}$	75	0.4	45	0.089	[41]
$Ba_{0.85}Ca_{0.15}Ti_{0.90}Hf_{0.10}O_3$	123	0.74	35	0.21	[42]
$Ba_{0.98}Ca_{0.02}Zr_{0.085}Ti_{0.915}O_3$	85	0.6	40	0.15	[43]
$Ba_{0.85}Ca_{0.15}Zr_{0.1}Ti_{0.895}Fe_{0.005}O_3$	72	0.60	33	0.182	[44]
$Gd_{0.02}Na_{0.5}Bi_{0.48}TiO_3$	97	0.75	90	0.083	[45]

8.4. FERROELECTRICS IN MAGNETOELECTRIC DEVICES

Coupling between electric and magnetic polarization enhances one more degree of freedom to control electric and magnetic field in the devices. The well-known multiferroic materials with the both the state is bismuth ferrite. It contains a spin glass state which makes it more promising. Hence advanced ceramics like $La_3Ni_2NbO_9$ with both ferroelectric nature as well as the spin glass state makes it very promising material for the magneto-electric coupling devices.

8.4.1. $La_3Ni_2NbO_9$

$La_3Ni_2NbO_9$ is a double perovskite with the structure of $A_3B_2CO_9$ type. Both the ferroelectric state as well as the spin glass state can co-exist in the ceramic structure as reported [46]. The disorder in the structure shown in Fig. (**8.2**) [46] is responsible for the coexistence of both the states.

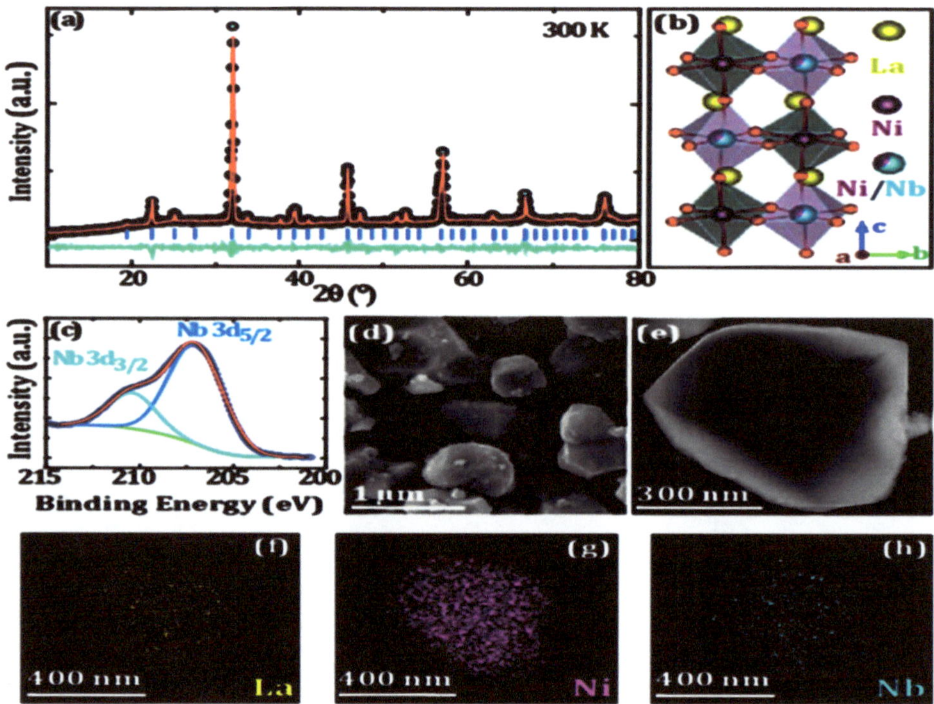

Fig. (8.2). (a) XRD **(b)** schematic diagram. **(c)** XPS **(d)** SEM **(e)** single particle and **(f-h)** element analysis of $La_3Ni_2NbO_9$. Reprint with the permission from Ref [46]. Copyright 2016, ACS Publishing.

8.5. FERROELECTRICS IN DRAM DEVICES

Dynamic random-access memory devices consisted with one capacitor and transistor together with. In the DRAM devices the role of the capacitor is to store the charge as the digital data whereas the role of the transistor is the selection of the memory cell. A range of materials such as aluminum oxide, zirconium oxide, hafnium oxide is in the use of the making of the DRAM devices. But a solid solution between the hafnium oxide and zirconium oxide can attribute a high k dielectric property for the application in the field.

8.5.1. (Hf, Zr)O$_2$

The presence of morphotropic boundary between the hafnium oxide and zirconium oxide resulted a high dielectric constant in the solid solution (Hf, Zr)O$_2$ as shown in Fig. (**8.3**) [47]. The main advantage of the ceramic is that with the decrease of the thickness the dielectric constant increases. This feature makes it one of the most promising materials for the application of the DRAM devices.

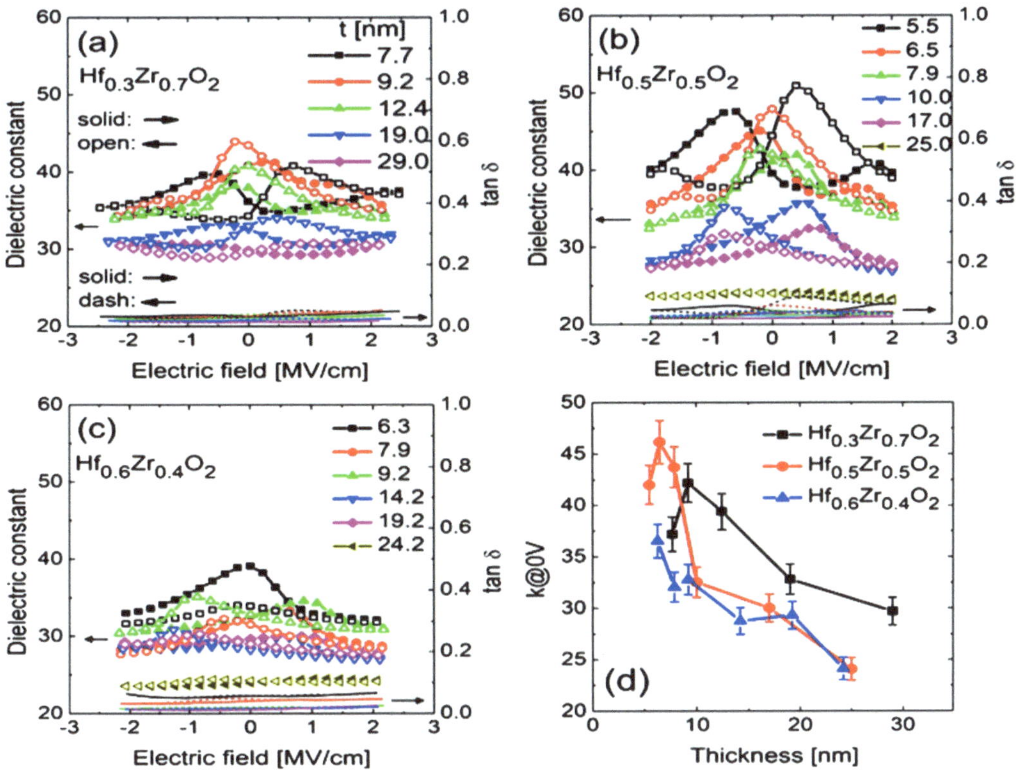

Fig. (8.3). (a-c) Dielectric constant *vs* electric field graphs and **(d)** loss *vs* electric field curves of (Hf, Zr)O$_2$

ceramics. Reprint with the permission from Ref [47]. Copyright 2018, ACS Publishing.

CONCLUSIONS

In summary, ferroelectric material contains intrinsic polarization which helps them to get polarized with the application of the electric field. Due to the presence of the retentivity and the coercive field they show a hysteresis loop pattern. But due to the presence of different artifacts the nature of such hysteresis loop need not to be symmetric in nature. In general, the non-centrosymmetric nature of the material such as the barium titanate makes these materials ferroelectric in nature. Hence, electrets which contain an artificial intrinsic polarization do not make it the ferroelectric in nature. Actually, the ferroelectrics are a class of materials comes under the dielectric materials. The other class of materials are pyroelectrics, piezoelectrics *etc*. In the earlier days of discovery of the ferroelectric materials it was believed to be the hydrogen bond responsible for the ferroelectric nature. But with the discovery of barium titanate, the non-centrosymmetric nature of the lattice structure starts coming into the picture. The present-day application of the ferroelectric materials ranges from the ferroelectric memory devices, electro caloric devices, magneto electric devices, DRAM capacitors *etc*., although, the lead zirconium titanate is the most popular material for the fabrication of nonvolatile ferroelectric random memory device. But the presence of lead makes it environmentally unfriendly. Hence, new class of materials like $Bi_{3.25}La_{0.75}Ti_3O_{12}$ is very promising for such applications. Not only that these advanced ceramics also possesses flexible properties along with the high writing speed and cyclability than that of the MoS_2 materials. Electro caloric devices on the other hand found application for the cooling of the microelectronics devices. Barium hafnium titanate prepared with the solid-state reaction technique showed large ECE values of 1.64 K for x=0.05 for these purposes. Presence of both spin glass state as well as the ferroelectricity makes it possible to couple both magnetic and electric field within the same material for the fabrication of the magneto-electric devices. Hence advanced ceramics like $La_3Ni_2NbO_9$ with both ferroelectrics' nature as well as the spin glass state makes it very promising material for the magneto-electric coupling devices. DRAM devices in general consist of one capacitor and transistor. In the DRAM devices the role of the capacitor is to store the charge as the digital data whereas the role of the transistor is the selection of the memory cell. Presence of morphotropic phase boundary in the advanced ceramics of hafnium oxide and zirconium oxide can result high dielectric constant for the DRAM applications. Hence, a range of advanced ceramics is available for the different device applications.

CONSENT FOR PUBLICATION

Not applicable.

CONFLICT OF INTEREST

The authors confirm that this chapter content has no conflict of interest.

ACKNOWLEDGEMENTS

The author would like to thank UGC, New Delhi for start-up grant no F.30-457/2018 (BSR). We also acknowledge the support provided to A. Franco Jr. by CNPq, Brazil with grant No. 307557/2015-4.

REFERENCES

[1]   P. Banerjee Jr, "A. F., Basha, D. B., & Naidu, K. C. B. Magnetic Nanomaterials for Spintronics. Magnetochemistry", *Materials and Applications,* vol. 66, p. 323, 2020.

[2]   A.K. Tagantsev, and G. Gerra, "Interface-induced phenomena in polarization response of ferroelectric thin films", *J. Appl. Phys.,* vol. 100, no. 5, 2006.051607
[http://dx.doi.org/10.1063/1.2337009]

[3]   K. Srinivas, K. C. B. Naidu, G. Balakrishna, B. V. S. Reddy, N. S. Kumar, and S. Ramesh, *Materials and Applications,* vol. 66, p. 259, 2020.

[4]   P. Banerjee, A.F. Junior, D.B. Basha, K.C.B. Naidu, and K. Srinivas, Niobium Based Materials for Supercapacitors. *Inorganic Nanomaterials for Supercapacitor Design.* CRC Press, 2019, pp. 1-15.

[5]   W.L. Warren, D. Dimos, G.E. Pike, K. Vanheusden, and R. Ramesh, "Alignment of defect dipoles in polycrystalline ferroelectrics", *Appl. Phys. Lett.,* vol. 67, no. 12, pp. 1689-1691, 1995.
[http://dx.doi.org/10.1063/1.115058]

[6]   P. Banerjee, A.F. Junior, D.B. Basha, K.C.B. Naidu, and S. Ramesh, Zinc-Based Materials for Supercapacitors. *Inorganic Nanomaterials for Supercapacitor Design.* CRC Press, 2019, pp. 17-31.
[http://dx.doi.org/10.1201/9780429277900-2]

[7]   H.V. Pessoni, T.E. Alves, P. Banerjee, and A.F. Júnior, "Functional properties of Ho3+ substituted cobalt ferrite in the context of the reduced mass model", *Physica B,* vol. 575, 2019.411676
[http://dx.doi.org/10.1016/j.physb.2019.411676]

[8]   U. Naresh, R.J. Kumar, S. Ramesh, K.C.B. Naidu, D.B. Basha, P. Banerjee, and K. Srinivas, *Conducting Polymer-Derived Materials for Batteries.* Conducting Polymers-Based Energy Storage Materials, 2019, pp. 65-78.

[9]   F.S. Oliveira, C.A.M. dos Santos, A.J.S. Machado, P. Banerjee, and A. Franco, "Structure and dielectric properties of Ba2CuxY1-xTaO6-y double perovskite", *SN Applied Sciences,* vol. 1, no. 11, p. 1447, 2019.
[http://dx.doi.org/10.1007/s42452-019-1479-z]

[10]   P. Banerjee, and A. Franco Jr, "Role of higher valent substituent on the dielectric and optical properties of Sr0.8Bi2.2Nb2O9 ceramics", *Mater. Chem. Phys.,* vol. 225, pp. 213-218, 2019.
[http://dx.doi.org/10.1016/j.matchemphys.2018.12.075]

[11]   F.I. Mopsik, and M.G. Broadhurst, "Molecular dipole electrets", *J. Appl. Phys.,* vol. 46, no. 10, pp. 4204-4208, 1975.
[http://dx.doi.org/10.1063/1.321433]

[12] H.O. Jacobs, and G.M. Whitesides, "Submicrometer patterning of charge in thin-film electrets", *Science,* vol. 291, no. 5509, pp. 1763-1766, 2001.
[http://dx.doi.org/10.1126/science.1057061] [PMID: 11230687]

[13] R. Kressmann, G.M. Sessler, and P. Gunther, "Space-charge electrets", *IEEE Trans. Dielectr. Electr. Insul.,* vol. 3, no. 5, pp. 607-623, 1996.
[http://dx.doi.org/10.1109/94.544184]

[14] G. Heiland, and H. Ibach, "Pyroelectricity of zinc oxide", *Solid State Commun.,* vol. 4, no. 7, pp. 353-356, 1996.
[http://dx.doi.org/10.1016/0038-1098(66)90187-6]

[15] C.C. Hsiao, and S.Y. Yu, "Rapid deposition process for zinc oxide film applications in pyroelectric devices", *Smart Mater. Struct.,* vol. 21, no. 10, 2012.105012
[http://dx.doi.org/10.1088/0964-1726/21/10/105012]

[16] D.L. Polla, R.S. Muller, and R.M. White, "Monolithic integrated zinc-oxide on silicon pyroelectric anemometer", *1983 International Electron Devices Meeting,* 1983pp. 639-642
[http://dx.doi.org/10.1109/IEDM.1983.190588]

[17] Y. Tan, J. Zhang, Y. Wu, C. Wang, V. Koval, B. Shi, H. Ye, R. McKinnon, G. Viola, and H. Yan, "Unfolding grain size effects in barium titanate ferroelectric ceramics", *Sci. Rep.,* vol. 5, p. 9953, 2015.
[http://dx.doi.org/10.1038/srep09953] [PMID: 25951408]

[18] Y. Luo, I. Szafraniak, N. D. Zakharov, V. Nagarajan, M. Steinhart, and R. B. Wehrspohn, *Appl. Phys. Lett.,* vol. 83, no. 3, pp. 440-442, 2003.
[http://dx.doi.org/10.1063/1.1592013]

[19] G.H. Kwei, A.C. Lawson, S.J.L. Billinge, and S.W. Cheong, "Structures of the ferroelectric phases of barium titanate", *J. Phys. Chem.,* vol. 97, no. 10, pp. 2368-2377, 1993.
[http://dx.doi.org/10.1021/j100112a043]

[20] W.S. Yun, J.J. Urban, Q. Gu, and H. Park, "Ferroelectric properties of individual barium titanate nanowires investigated by scanned probe microscopy", *Nano Lett.,* vol. 2, no. 5, pp. 447-450, 2002.
[http://dx.doi.org/10.1021/nl015702g]

[21] H. Kohlstedt, Y. Mustafa, A. Gerber, A. Petraru, M. Fitsilis, and R. Meyer, *Microelectron. Eng.,* vol. 80, pp. 296-304, 2005.
[http://dx.doi.org/10.1016/j.mee.2005.04.084]

[22] K. Kim, and Y.J. Song, "Integration technology for ferroelectric memory devices", *Microelectron. Reliab.,* vol. 43, no. 3, pp. 385-398, 2003.
[http://dx.doi.org/10.1016/S0026-2714(02)00285-8]

[23] R.I. Epstein, and K.J. Malloy, "Electrocaloric devices based on thin-film heat switches", *J. Appl. Phys.,* vol. 106, no. 6, 2009.064509
[http://dx.doi.org/10.1063/1.3190559]

[24] H. Gu, X. Qian, X. Li, B. Craven, W. Zhu, and A. Cheng, *Appl. Phys. Lett.,* vol. 102, no. 12, 2013.122904
[http://dx.doi.org/10.1063/1.4799283]

[25] S. Fusil, V. Garcia, A. Barthélémy, and M. Bibes, "Magnetoelectric devices for spintronics", *Annu. Rev. Mater. Res.,* vol. 44, pp. 91-116, 2014.
[http://dx.doi.org/10.1146/annurev-matsci-070813-113315]

[26] S.H. Baek, H.W. Jang, C.M. Folkman, Y.L. Li, B. Winchester, J.X. Zhang, Q. He, Y.H. Chu, C.T. Nelson, M.S. Rzchowski, X.Q. Pan, R. Ramesh, L.Q. Chen, and C.B. Eom, "Ferroelastic switching for nanoscale non-volatile magnetoelectric devices", *Nat. Mater.,* vol. 9, no. 4, pp. 309-314, 2010.
[http://dx.doi.org/10.1038/nmat2703] [PMID: 20190772]

[27] S. K. Kim, G. J. Choi, S. Y. Lee, M. Seo, S. W. Lee, and J. H. Han, *Adv. Mater.,* vol. 20, no. 8, pp.

1429-1435, 2008.
[http://dx.doi.org/10.1002/adma.200701085]

[28] D.E. Kotecki, "A review of high dielectric materials for DRAM capacitors", *Integr. Ferroelectr.,* vol. 16, no. 1-4, pp. 1-19, 1997.
[http://dx.doi.org/10.1080/10584589708013025]

[29] H.P. Baltes, and R.S. Popovic, "Integrated semiconductor magnetic field sensors", *Proc. IEEE,* vol. 74, no. 8, pp. 1107-1132, 1986.
[http://dx.doi.org/10.1109/PROC.1986.13597]

[30] P. Ripka, and M. Janosek, "Advances in magnetic field sensors", *IEEE Sens. J.,* vol. 10, no. 6, pp. 1108-1116, 2010.
[http://dx.doi.org/10.1109/JSEN.2010.2043429]

[31] G. Rosenman, and I. Rez, "Electron emission from ferroelectric materials", *J. Appl. Phys.,* vol. 73, no. 4, pp. 1904-1908, 1993.
[http://dx.doi.org/10.1063/1.354059]

[32] H. Riege, "Electron emission from ferroelectrics-a Review", *Nucl. Instrum. Methods Phys. Res. A,* vol. 340, no. 1, pp. 80-89, 1994.
[http://dx.doi.org/10.1016/0168-9002(94)91282-3]

[33] A. Forbes, A. Dudley, and M. McLaren, "Creation and detection of optical modes with spatial light modulators", *Adv. Opt. Photonics,* vol. 8, no. 2, pp. 200-227, 2016.
[http://dx.doi.org/10.1364/AOP.8.000200]

[34] A.P. Kowalczyk, M. Makowski, I. Ducin, M. Sypek, and A. Kolodziejczyk, "Collective matrix of spatial light modulators for increased resolution in holographic image projection", *Opt. Express,* vol. 26, no. 13, pp. 17158-17169, 2018.
[http://dx.doi.org/10.1364/OE.26.017158] [PMID: 30119531]

[35] L. Su, X. Lu, L. Chen, Y. Wang, G. Yuan, and J.M. Liu, "Flexible, fatigue-free, and large-scale Bi3.25La0.75Ti3O12 ferroelectric memories", *ACS Appl. Mater. Interfaces,* vol. 10, no. 25, pp. 21428-21433, 2018.
[http://dx.doi.org/10.1021/acsami.8b04781] [PMID: 29863844]

[36] M.D. Li, X.G. Tang, S.M. Zeng, Q.X. Liu, Y.P. Jiang, T.F. Zhang, and W.H. Li, "Large Electrocaloric Effect in Lead-free Ba (HfxTi1–x)O3 Ferroelectric Ceramics for Clean Energy Applications", *ACS Sustain. Chem.& Eng.,* vol. 6, no. 7, pp. 8920-8925, 2018.
[http://dx.doi.org/10.1021/acssuschemeng.8b01277]

[37] F. Han, Y. Bai, L.J. Qiao, and D. Guo, "A systematic modification of the large electrocaloric effect within a broad temperature range in rare-earth doped BaTiO3 ceramics", *J. Mater. Chem. C Mater. Opt. Electron. Devices,* vol. 4, no. 9, pp. 1842-1849, 2016.
[http://dx.doi.org/10.1039/C5TC04209G]

[38] S.K. Upadhyay, V.R. Reddy, P. Bag, R. Rawat, S.M. Gupta, and A. Gupta, "Electro-caloric effect in lead-free Sn doped BaTiO3 ceramics at room temperature and low applied fields", *Appl. Phys. Lett.,* vol. 105, no. 11, 2014.112907
[http://dx.doi.org/10.1063/1.4896044]

[39] Y. Bai, X. Han, K. Ding, and L-J. Qiao, "Combined effects of diffuse phase transition and microstructure on the electrocaloric effect in Ba1-xSrxTiO$_3$ ceramics", *Appl. Phys. Lett.,* vol. 103, 2013.162902
[http://dx.doi.org/10.1063/1.4825266]

[40] Z. Luo, D. Zhang, Y. Liu, D. Zhou, Y. Yao, C. Liu, B. Dkhil, X. Ren, and X. Lou, "Enhanced electrocaloric effect in lead-free BaTi1-xSnxO$_3$ ceramics near room temperature", *Appl. Phys. Lett.,* vol. 105, 2014.102904
[http://dx.doi.org/10.1063/1.4895615]

[41]   Y. Zhao, X.Q. Liu, J.W. Wu, S.Y. Wu, and X.M. Chen, "Electrocaloric effect in relaxor ferroelectric $Ba(Ti_{1-x}Y_x)O_{3-x/2}$ ceramics over a broad temperature range", *J. Alloys Compd.,* vol. 729, pp. 57-63, 2017.
[http://dx.doi.org/10.1016/j.jallcom.2017.09.161]

[42]   X. Wang, J. Wu, B. Dkhil, C. Zhao, T. Li, W. Li, and X. Lou, "Large electrocaloric strength and broad electrocaloric temperature span in lead-free $Ba_{0.85}Ca_{0.15}Ti_{1-x}Hf_xO_3$ ceramics", *RSC Advances,* vol. 7, pp. 5813-5820, 2017.
[http://dx.doi.org/10.1039/C6RA27628H]

[43]   J. Wang, T. Yang, S. Chen, G. Li, Q. Zhang, and X. Yao, "Nonadiabatic direct measurement electrocaloric effect in lead-free $Ba,Ca(Zr, Ti)O_3$ ceramics", *J. Alloys Compd.,* vol. 550, pp. 561-563, 2013.
[http://dx.doi.org/10.1016/j.jallcom.2012.10.144]

[44]   S. Patel, A. Chauhan, and R. Vaish, "Enhanced electrocaloric effect in Fe-doped $(Ba_{0.85}Ca_{0.15}Zr_{0.1}Ti_{0.9})O_3$ ferroelectric ceramics", *Appl.Mater. Today,* vol. 1, pp. 37-44, 2015.
[http://dx.doi.org/10.1016/j.apmt.2015.08.002]

[45]   M. Zannen, A. Lahmar, Z. Kutnjak, J. Belhadi, H. Khemakhem, and M. El Marssi, "Electrocaloric effect and energy storage in lead free $Gd_{0.02}Na_{0.5}Bi_{0.48}TiO_3$ ceramic", *Solid State Sci.,* vol. 66, pp. 31-37, 2017.
[http://dx.doi.org/10.1016/j.solidstatesciences.2017.02.007]

[46]   K. Dey, A. Indra, D. De, S. Majumdar, and S. Giri, "Magnetoelectric coupling, ferroelectricity, and magnetic memory effect in double perovskite $La_3Ni_2NbO_9$", *ACS Appl. Mater. Interfaces,* vol. 8, no. 20, pp. 12901-12907, 2016.
[http://dx.doi.org/10.1021/acsami.6b02990] [PMID: 27136317]

[47]   M.H. Park, Y.H. Lee, H.J. Kim, Y.J. Kim, T. Moon, K.D. Kim, S.D. Hyun, and C.S. Hwang, "Morphotropic phase boundary of $Hf_{1-x}Zr_xO_2$ thin films for dynamic random access memories", *ACS Appl. Mater. Interfaces,* vol. 10, no. 49, pp. 42666-42673, 2018.
[http://dx.doi.org/10.1021/acsami.8b15576] [PMID: 30468068]

CHAPTER 9

Transport Properties of Semiconducting Glasses: A Review

K. V. Ramesh[1,*] and **N.V. Krishna Prasad**[2]

[1] *Department of Electronics and Physics, GITAM Deemed to be University, Rushikonda, Visakhapatnam- 530045, Andhra Pradesh, India*

[2] *Dept. of Physics, GITAM Deemed to be University, Bangalore-562163, Karnataka, India*

Abstract: Research related to amorphous semiconductors in thin films has gained tremendous significance due to their applications in potential areas. Lead vanadate that belongs to the category of semiconducting oxide glasses has been studied to the extent of consideration. Apart from their application point of view, it is necessary to understand the physical properties of these materials. Taking into consideration wide variation in material composition, it is evident that large scope prevails in terms of research. Many laboratories were carrying extensive research on binary, tertiary mixed metal oxide glasses along with their derivatives/substitution of other metal oxides or transition of rare earth metal oxides was identified to suit for glass system showing considerable variation in physical properties if substituted for different applications. A review on these glass systems has been reported in this paper in particular transport properties of semiconducting glasses.

Keywords: Ceramics, Metal Oxides, Semiconducting Glasses, Transition, Transportation.

9.1. INTRODUCTION

The study of amorphous materials is very interesting because of their peculiar character in contrast to the crystalline materials. In recent times more emphasis on the scientific study of these materials increased because of their technological importance or promise. Though a large number of investigations have been made on these non-crystalline materials, a large part of the physical properties is still to be understood. The wave-like nature of the electron in a periodic potential of a crystalline lattice was invoked to develop the band theory for better awareness related to their physical properties.

---

* **Corresponding author Prof. K V Ramesh:** GITAM Deemed to be University, Rushikonda, Visakhapatnam-530045, A.P, India; E-mail: kvramesh11@gmail.com

But the absence of the long-range periodicity in amorphous materials makes this task rather complicated. In amorphous materials, atomic arrangement do not have periodicity (long range order) found in the crystalline materials. However short range order exists. These materials possess randomness which can exist in different forms like spin, topological, vibrational or substitutional disorder. (Fig. **9.1**) [1] shows the topological disorder. Amorphous materials that exhibit the phenomenon of glass transition are referred to as glasses. A material formed by cooling from molten state and exhibiting zero discontinuous changes in first order thermodynamic parameters is known as glass.

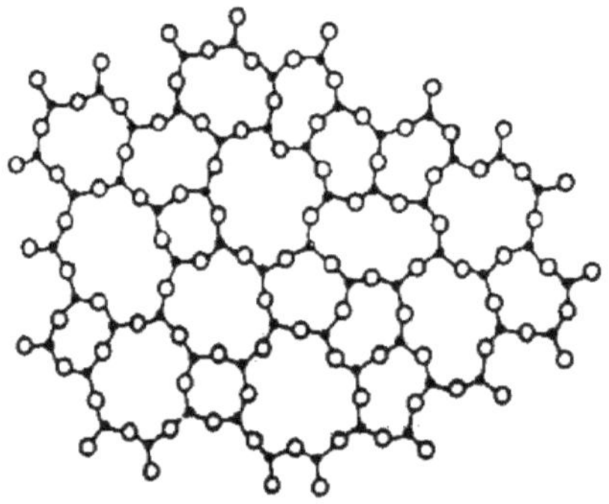

Fig. (9.1). Two dimensional continuous random networks as proposed by Zachariasen (taken from S.R. Elliot) [1].

9.2. GLASS TRANSITION

The transition from a liquid to glassy state is denoted by glass transition. In the liquid state a substance flows under any shear stress and its structure a change with changing temperature. In the glassy state, the substance, a glass, is solid. It does not flow irreversibly, and the structure is frozen. A glass has a molecular structure with order in a short range (a few molecular dimensions), but no long-range order at many molecular distances. (Fig. **9.2**) [2] shows schematically the change in volume V, enthalpy H, or entropy S of a liquid during cooling. For substances which can exist in both glassy and crystalline states, there are two possible paths below the melting temperature T_m. One of these paths is the usual one for crystalline solids below T_m and the other is a super-cooled liquid below T_m.

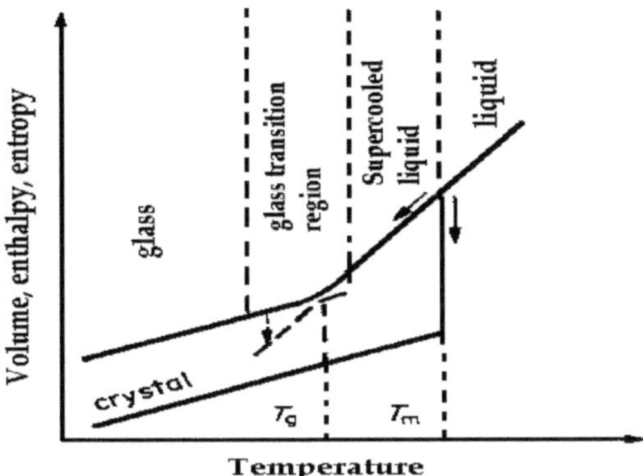

Fig. (9.2). Temperature dependence of properties of liquid, glass and crystal.

$$\partial G / \partial T = -S \tag{9.1}$$

where G is the Gibbs free energy.

It follows from (Fig. **9.3**) [3] that to form a glass, the liquid must make a thermodynamically un-favorable choice and follow the path with higher free energy, for kinetic reasons. At the temperature T_m, crystallization is impeded by a finite viscosity of the liquid. By rapid cooling, crystallization is kinetically arrested. Inorganic glass forming melts have a high viscosity at T_m and can be cooled slowly. In contrast, molten metals require cooling rates of the order of 10^5 K s^{-1} to form glass. An expression relating T_g and cooling rate derived from free-volume theory is given by:

$$q = q_o \ \exp\left[-\tfrac{1}{C}\left(1/T_g - 1/T_m\right)\right] \tag{9.2}$$

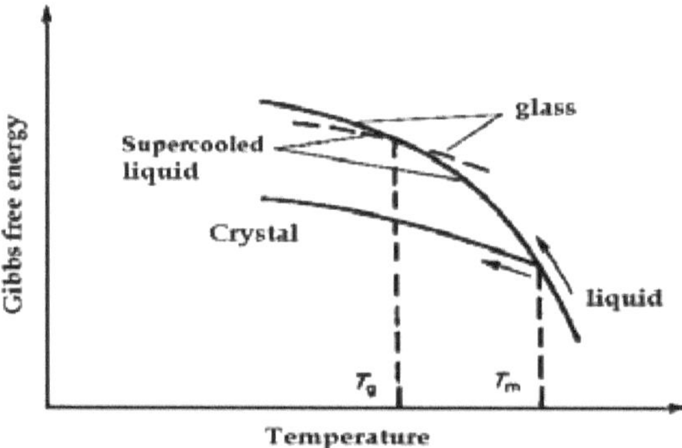

Fig. (9.3). Temperature dependence of the Gibbs free energy for liquid, glass and crystal.

where C is a constant ($\sim 3 \times 10^{-5}$) and for chalcogenide glasses q_0 varies between 10^{23} K s^{-1} and 10^4 K s^{-1} [2]. A super-cooled liquid is in metastable equilibrium with respect to a crystalline solid. The free energy profile can be pictured as one with a minimum for a crystalline state and a "trough" above the minimum, for the super-cooled liquid state. Below the value of T_g, the glass structure is not in equilibrium. It is frozen in by a viscosity $10^{15} - 10^{20}$ times greater than that of ordinary liquids. The liquid is never completely frozen, even at temperature below T_g its volume decreases slowly in the direction of its equilibrium value. This process is called stabilization and evolves on a geological time scale.

9.3. LOCALIZATION IN AMORPHOUS MATERIALS

For understanding the electrical properties, electron energy band theory of a crystalline solid can be explained systematically because of the presence of long-range periodicity. The state of electron is denoted by Bloch function Ψ. But in the case of amorphous or disordered solids such systematic explanation for electrical properties is not possible. In amorphous materials sufficient disorder can be produced and the electrons may become localized. Density of states is the concept that carried first from theory of crystals to theory of amorphous materials. In amorphous solids the band gap exists, but in a modified form due to disorder. The edges of valance and conduction bands enter into band gap and may overlap. The conductivity of amorphous materials essentially comes from these localized states or extended states.

The concept of localization in random lattices or amorphous materials was first addressed by Anderson [3] by considering tight binding approximation. There may be two types of disorders in a solid (i) absence of long-range order (ii) vertical fluctuation of potential with a specific range (V_o) at each potential well. The solution of Schrodinger equation then shows that there will be a random fluctuation of ψ for a pair of electronic wave function. These states are called non-localized. But if V_o/B, B being the bandwidth becomes larger than a certain value then the wave function falls off as exp (-2αR) and localization occurs (Fig. **9.4**) [3]. α^{-1} is called the localization length. However, Anderson condition is not unique and depends on the coordination number Z. The localization criterion may be rewritten in the general form B/ZV_o, so that localization on lattices having different values of coordination number Z, may be compared. The measure of degree of localization is given by α^{-1}. Smaller the value of α^{-1}, greater is the degree of localization. Another measure of the degree of localization that can be used is the "participation ratio" P. The P value denotes sites with electronic wave functions having sufficient amplitude. Obviously, the condition $P \rightarrow 0$ is another criterion for strong localization. Since the conduction between localized states can take place only due to the thermally assisted 'hopping', for states with energy E, the DC conductivity's ensemble average $<\sigma_E>$ at absolute zero must be zero.

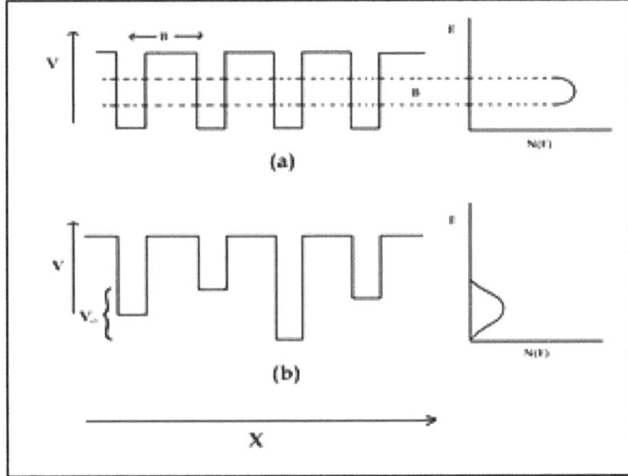

Fig. (9.4). (a) Potential well for a crystalline lattice and **(b)** Potential well for the Anderson lattice and the density of states N(E) is also shown.

One consequence of the long-range disorder of the amorphous solids is localization. Another consequence is the 'tailing' of states into the gap at the band edges as shown in Fig. **(9.5)** [4]. Bond angle distortion or fluctuation of short-range order causes broadening of bands and subsequent tailing of states into the band gap. A finite probability exists for the electron to be in the band tail. σ_E will be finite in the middle of the band if Anderson criteria are not satisfied then, but zero for energies.

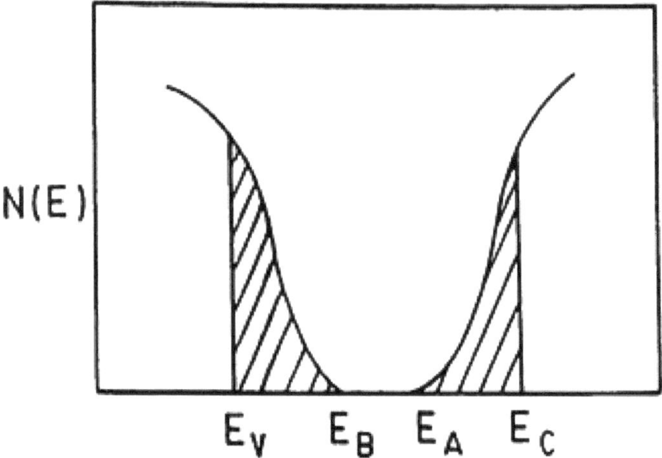

Fig. (9.5). The tailing of states (shaded) into the band gap at the amorphous material with continuous random network without defects.

near its extremities. Mott [4, 5] pointed out that a critical energy E_c must separate localized and non-localized states, and defined as:

$$<\sigma_E> = 0 \quad E < E_c$$
$$\neq 0 \quad E > E_c$$

This is illustrated for a density of states resulting from the Anderson potential in (Fig. **9.6**) [5]. E_c and E_c^1 separate localized and non-localized states at the bottom and top edge of the band respectively. The electronic wave function in the extended states ranged to infinity and the electron can travel from one end to the other of the material, giving a non-zero $<\sigma_E>$ in the limit $T \rightarrow 0$ while the wave function exponentially decays with distance for localized electron and hence contributes zero conductivity in the limit $T \rightarrow 0$. So, the mobilities in the localized and non-localized states are quite different. The energies E_c and E_c^1 which separates these two types of states are called "mobility edges". The density of states N(E) and its derivatives are continuous at E_c and E_c^1.

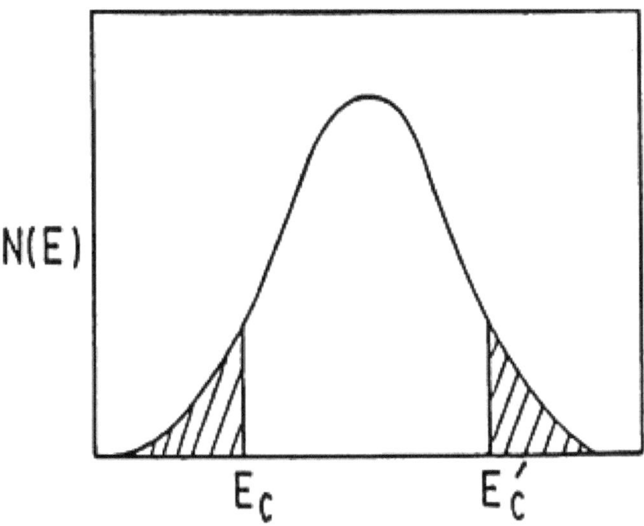

Fig. (9.6). Density of states in the Anderson model when states are nonlocalized in the centre of the band. Localized states are shown shaded. E_c, E'_c separate the range of energy where states are localized and nonlocalized.

9.4. MODELS FOR BAND THEORY IN AMORPHOUS MATERIALS

There are various band models for amorphous materials. As previously mentioned, density of electron states validates the concept for non-crystalline materials. Since the short order range is preserved in materials of non-crystalline type and in the case of crystalline materials the total features remains unchanged in terms of density of states. To calculate the density of states with emphasis of short-range order new techniques are proposed [6]. The proportionality between $N(E)$ and $E^{1/2}$ obtained for crystals is not applicable for non-crystalline solids [7].

9.4.1. CFO (Cohen, Frizsche and Ovshinsky) Model

The question of localized states is of considerable importance because of their contribution to the conductivity of the materials. Cohen, Frizsche and Ovshinsky [8] assumed that the structure belonging to non-crystalline type may lead to overlap of localized states band tails as shown in Fig. (**9.7a**) [8].

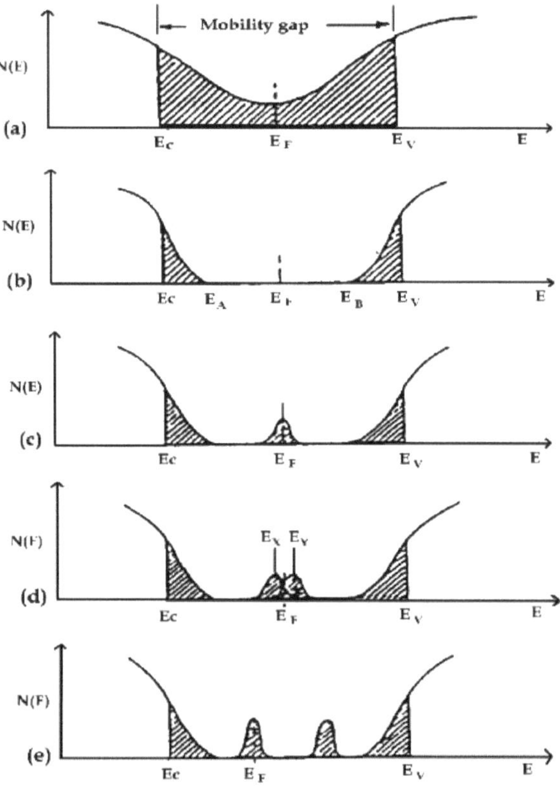

Fig. 9.7. Various forms proposed for the density of states in amorphous semi-conductors. Localized states are shown shaded, **(a)** Mott – CFO Model; **(b)** A real gap in the density of states suggested for random network without defects; **(c)** Davis and Mott model; **(d)** the same as **(c)** but with over lapping bands of donors (E_x) and acceptors (E_y) levels arising from defects; **(e)** Marshall and Owen model.

Similarly, those derived from conduction band will be neutral when vacant and those from the valence band neutral when filled. In the overlapping region they will be charged leading to centers with unpaired electron spin. Such overlapping states may pin the Fermi energy. Another feature of this model is the presence of mobility edges. These are identified as energies separating localized from non-localized states as was introduced by Mott [4, 5], therefore this model in some cases called as Mott - CFO model. The difference of energies between mobility edges of valance (E_v) and conduction bands (E_c) is called the "mobility gap". Even though evidence exists for the presence of mobility edges, the overlapping of tails seems to be invalid to apply to amorphous semiconductors and insulators, which are transparent to visible or infrared regions. It is believed that amorphous semiconductors in which all bands are saturated and in which there is no long-range fluctuation should have a band as shown in Fig. (**9.7b**) [8]. The overlapping tails exists in some liquids, such as expanded fluid mercury, which undergo metal - insulator transition under change in volume or temperature.

9.4.2. Davis and Mott Model

An alternative suggestion for band model was given by Davis and Mott [9]. On the basis of several experimental results they proposed the existence of finite density of states Fig. **(9.7c)** [8] at the Fermi energy. In this model electrons originating from a weaker band of donars occupy band of deep acceptors partially. Also donors and acceptors and their roles can be reversed. The model allows optical transparency as long as the total density of states in the gap is not large and with no need of assumptions regarding the magnitude of the matrix elements. However, no explanation was given to explain why the controlling states should lie near mid-gap. As per suggestions of Mott [10], if the states aroused from defect centre (dangling bonds, interstitials *etc.*), they act both as deep donors (E_y) and acceptors (E_x), single and double occupancy conditions leading to two bands separated by an appropriate correlation energy or "Hubbard U" as shown in Fig. **(9.7d)** [8].

9.4.3. Marshall - Owen model

Model was suggested by Marshall and Owen [11] is based on the observation of field effect study on As_2Te_3 and high field drift mobility in $As_2\ Se_3$. Bands of donors and acceptors in the upper and lower region of the gap determine the position of Fermi level. The concentration levels of donors and acceptors get adjusted by their own self-compensation to nearly equal such that the Fermi level stays near the center of the gap. At lower temperatures it moves towards any one impurity bands as shown in Fig. **(9.7e)** [8].

9.4.4. D^+D^- model

This model was proposed by Street and Mott [12] Mott, Davis and Street [13] to explain the absence of ESR and Para magnetism of Chalcogenide glasses. It is supposed that an unoccupied dangling bond forms a covalent bond with the lone pair orbital on a neighboring chalcogenide atom and thus has a large binding energy. If D^+, D° and D^- denotes the dangling bonds with zero, one and two electrons, the reaction:

$$2D^\circ \rightarrow D^+ + D^-$$

is assumed to be exothermic. Thus, all the dangling bonds are positively or negatively charged. Fermi energy is determined by the energies of the sites D^+ and D^- and there are no free spins. The broadening due to disorder is small and so localized states do not overlap.

POLARON

For strong localization, the conduction is essentially due to thermally activated hopping of electron between the localized states in the mobility gap. This strong localization induces the formation of polaron as a conducting species. Also it can be assessed that a charge carrier in one state always twist the surroundings, and this twist will be significant for more localized carrier, which in turn lowers the carrier energy leading to self trapping in extreme cases forming a polaron. Its effective mass is larger than that of Bloch electron. The size of a polaron is measured by the extent of the lattice distortion induced by the excess electron. For a non-crystalline material, states in the mobility gap are localized, hence the electrons are trapped, and cause distortion to the site producing polarons. Polaron is called "small polaron" if the potential of the carrier is due to single atom. In a material if the electron is self-trapped within distance r_p, called the polaron radius then its potential well is described in Fig. (**9.8**) by Mott *et al.* [14].

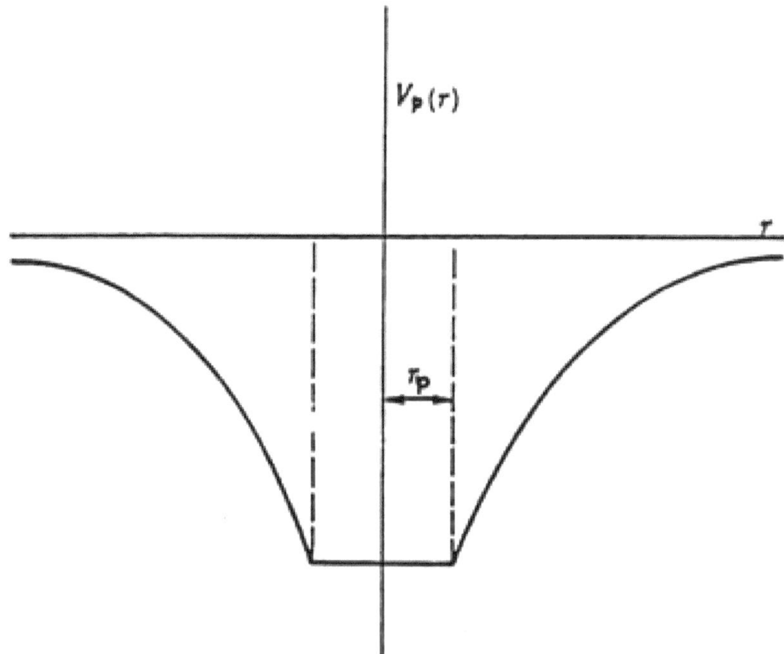

Fig. (9.8). Potential well due to the polarization of an ionic lattice round a trapped electron.

$$V_p(r) = \frac{-e^2}{\epsilon_p\, r} \quad (\text{where } r > r_p) \tag{9.3}$$

$$= \frac{-e^2}{\epsilon_p\, r_p} \quad (\text{where } r < r_p) \tag{9.4}$$

where $\dfrac{1}{\epsilon_p} = \dfrac{1}{\epsilon_\infty} - \dfrac{1}{\epsilon_o}$. Here ϵ_o and ϵ_∞ are static, high frequency dielectric constants respectively. The total energy of the self-trapped electron is minimized or lowered to find the r_p. The energy of the polaron was calculated to a first approximation [14] as

$$W_p = \frac{\pi^2 \hbar^2}{2m^* r_p^2} - \frac{e^2}{2\,\epsilon_p\, r_p} \tag{9.5}$$

where the 2nd term is the energy lowered to self-trapping and m represents effective mass in the undisturbed lattice. Minimizing the quantity in Eq. (12. 5) gives

$$r_p = \frac{2\pi^2 \hbar^2\, \epsilon_p}{m^* e^2} \tag{9.6}$$

and

$$W_p = \frac{e^2}{4\,\epsilon_p\, r_p} \tag{9.7}$$

If R the distance of electron to be transferred is not large compared to r_p the Eq. (12.7) for W_p is not valid, but is given by:

$$W_p = \frac{e^2}{2\,\epsilon_p}\left(\frac{1}{r_p} - \frac{1}{R}\right)$$

(9.8)

W_p as function of site separation R is shown in Fig. (**9.9**) [14].

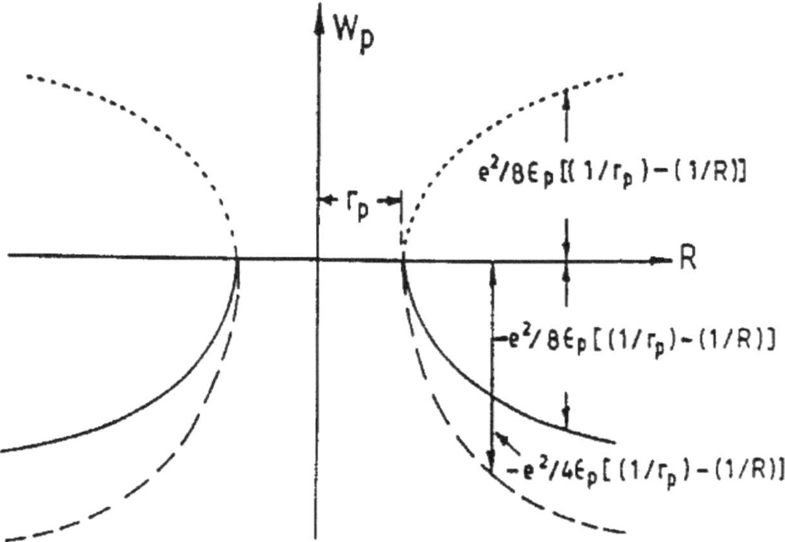

Fig. (9.9). Variation of effective binding energy W_p for a pair of sites as function of R. Dotted lines indicate lattice polarization energy and dashed line indicate potential energy of electron separating the two lines.

9.5. AMORPHOUS MATERIALS - PREPARATION

There exist many techniques for the preparation of amorphous materials [15] of which some are briefly discussed below.

9.5.1. Thermal Evaporation

This technique is understood in conceptual way easily. The material is taken in a boat, which can be heated electrically. The substrate is held above the boat, which can be preheated before the deposition starts. The whole arrangement is taken in a glass jar, which can be evacuated. The material is evaporated at low pressure $\sim 10^{-6}$ Torr and is deposited on the substrate giving an amorphous film.

9.5.2. Sputtering

The spluttering process is more complicated than thermal evaporation. In this process an r.f field is applied between the substrate and target. During one half of the r.f. cycle positive ions strike the target and during the other half cycle electrons strike the target. Electrons are more mobile than the positive ions and after a certain time electron are accumulated on the target. After that time these electrons attract positive ions from the plasma essentially by the bias potential, rather than the r.f. potential. The ejected material from the target is then carried to the substrate to be deposited in the form of an amorphous film.

9.5.3. Glow Discharge Decomposition

A technique similar to sputtering where chemical decomposition of gas takes place by its own leading to solid film deposition on a substrate placed in the plasma instead of the plasma ejecting material from a target. The production of plasma is by applying an r.f field with the help of inductive coupling or either capacitive coupling. Controlled doping is a major advantage of this technique.

9.5.4. Chemical Vapour Deposition

This technique is similar to glow discharge method. The only difference between these two is that here decomposition of reactant is caused by heating and r. f field is used to heat up the substrate upon which the vapour decomposes.

9.5.5. Melt Quenching

Melt quenching is the oldest technique to produce a glass. The most significant feature of melt quench method to produce an amorphous material is hardening continously, whereas crystallization is a process of discontinuous solidification as shown in Fig. (**9.10**) [25]. It is essential to cool the melt very fast during glass formation .The cooling rate of the melt should be higher than $1/\tau$ (where τ is the molecular relaxation time) [16]. The procedure for preparing glass in this methodology is very simple. The sample is taken in a pure alumina or silica or platinum crucible and heated in an electrical furnace. The molten mass is then quenched either on a copper twin roller or by pressing the melt between brass/steel plates to obtain the glass. By this method cooling rates of 10^{-2} - 10^{3} Ks^{-1} are achievable.

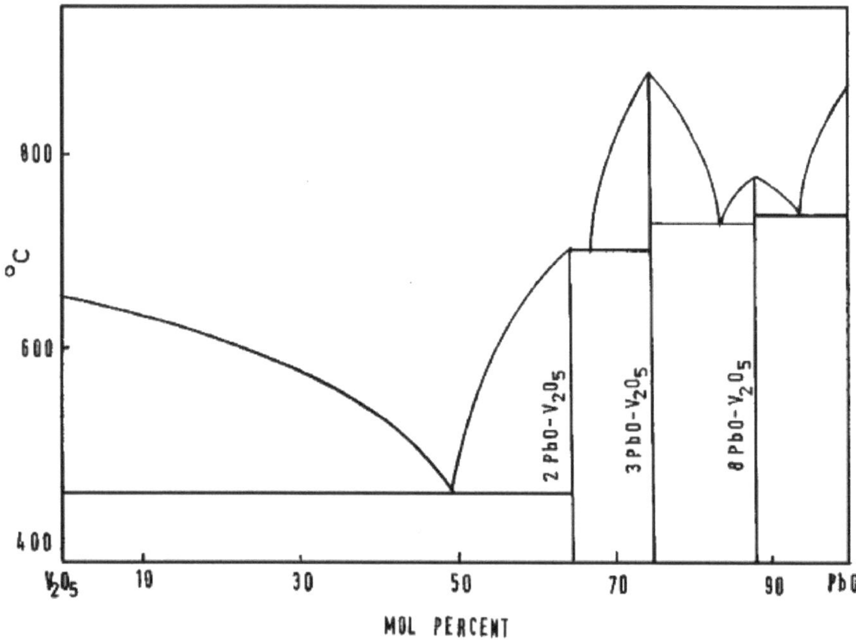

Fig. (9.10). Amadori's phase diagram of the PbO – V$_2$O$_5$ system.

Electrical conduction occurs due to ionic transport in the conventional silicate-based glasses while it is due to the transport of electrons in semi-conducting glasses. Large amount of research in analyzing physical and structural properties of glasses in ordinary and semi-conducting glasses has widely enhanced in view of their potential towards application of semi-conducting glasses. These applications include electrochemical batteries [19], threshold switching [17] and memory switching [18] *etc*. Research was carried on bulk semi-conducting glasses as well as thin or thick film forms.

Semi-conducting glasses are classified into [20] chalcogenide glasses and transition metal oxide glasses. Chalcogenide glasses consist of S or Se or Te elements compounded with III or IV or V group elements (As_2S_3 or As_2Se_3 *etc*.) while transition metal oxide glasses are formed by transition metal oxides like V_2O_5 or Fe_2O_3 or MoO_3 or WO_3 as major constituent. Glass formation in Vanadate glasses was reported by Denton *et al* [21] and Baynton *et al*. [22] reported their semiconducting nature. The phase diagram of PbO-V$_2$O$_5$ binary solid solution has been formulated by Amadori *et al*. as early as in 1917 [23]. The phase diagram shown in Fig. (**9.10**) indicates that 1:1 molar ratio is the eutectic composition and also this compound can be made into glass over a wide composition range (*i.e.*,

46-68 mole %) of V_2O_5 [21]. Minor modifications of this phase diagram has been done by Gregoir *et al.* [24] which include the existence of two eutectic compositions about the 1:1 molar ratio. Neverthless, the original phase diagram of Amadori *et al.* [23] is followed as the standard in literature and in present studies.

Baiocchi *et al.* studied the new phases formed in the PbO-V_2O_5 equimolar system depending on the cooling rate which was determined by the type of material on which the melt was quenched [25]. When the melt was quenched on graphite block, they obtained perfect glass pattern. When the melt was quenched on copper block, they obtained a glass sample (a sampled named MMl) which partially devitrified containing a crystalline phase of $Pb_2V_2O_7$ which is called α-LPV. Similarly, when the melt was quenched on stainless steel block, they obtained a glass sample (a sample named MM2) which also partially devitrified containing a different crystalline phase of $Pb_2V_2O_7$ which is called β-LPV. However, after annealing, α-LPV and β-LPV phases converted to lead metavanadate which was considered to be the most stable phase. This work was extended by Calestani *et al.* [26] who not only studied the 1:1 composition, but also higher V_2O_5/PbO molar ratios. They have reported the observation of two meta-stable phases of lead metavanadate named PbV_2O_6 (II) and PbV_2O_6 (III) in addition to the LPV phases reported by Baiocchi *et al.* [25], earlier. After heat treatment, all the meta-stable phases were found to be converting to the most stable phase of lead metavanadate named PbV_2O_6 (I). The lead metavanadate phase in glass and crystalline forms was studied from measurements of magnetic susceptibility [28], thermodynamic [26] and Electron Spin Resonance (ESR) [27] measurements.

Jordon and Calvo determined the crystal structure of PbV_2O_6 in its most stable form (phase I) by X-ray diffraction technique [29]. This compound belongs to orthorhombic space group Pnma with a = 9.771 Å, b = 3.68 Å and c = 12.713 Å with Z (no. of molecules per unit cell) = 4. The crystal structure is characterized by two unique vanadium ions which lie on mirror planes and are each coordinated to five oxygen atoms. These V-O distances lie between 1.6 - 2.06 Å. The distorted octahedra are completed by sixth oxygen interacting with vanadium. This V-O distance is 2.73 Å for V(1) and 2.57 Å for V(2). The lead ion is coordinated to nine oxygen atoms lying in a spherical shell. The structure consists of chains of VO_6 octahedra parallel to b-axis. The crystal structures of the other two meta-stable phases of PbV_2O_6 (II) and PbV_2O_6 (III) have also been determined [30, 31]. These two phases have densities 5.30 gm/cm^3 and 5.11 gm/cm^3 and co-ordination around vanadium is square pyramidal and tetrahedral respectively. The oxygen co-ordination around lead is eight in both the phases. X-ray diffraction studies were also carried out on the amorphous PbV_2O_6 by means of radial distribution function (RDF) analysis [32]. The RDF studies indicate that PbV_2O_6 glass structure cannot strictly be related to any of the known crystalline phases. The average coordination number n(V-O) around vanadium is 5 and the average vanadium oxygen distance r(V-O) is 1.73 Å. Similarly, the average value for n(Pb-O) is 6 and the average r(Pb-O) is 2.54 Å. These observations are in agreement with the results obtained by Wright *et al.* [33] in neutron diffraction studies on the vanadate glass systems of similar composition. Apart from equimolar compound many studies exist regarding glass systems formed with different molar ratios of PbO/V_2O_5. IR spectral studies on a number of $MnOm$-V_2O_5 glass systems including PbO - V_2O_5 with 20 and 50 mole % for PbO [34] were carried by Dimitriev *et al.*,. They postulated two different forms of metal ion distribution in the glasses in relation to vanadium - oxygen polyhedral from IR spectra according to which Pb^{2+}ion occupy interstitial position like other divalent ions Zn^{2+}, Cd^{2+} *etc.* PbO-V_2O_5 glass system in 0 - 75 mole % range for PbO was studied by Dimitrov and Dimitriev [34] who identified the basic structural units as affected VO_5 groups, unaffected VO_5 groups, V_2O_7 pyrovanadate units and VO_4 isolated tetrahedral units based on composition.

DTA, ESR and IR studies on $xPbO$-$(100-x)V_2O_5$ glass system in which PbO was varies from 10% to 50 mole % was carried out by Mandal and Ghosh [36]. Their outputs are in accordance with values of literature and hyperfine structure in ESR spectrum disappears at molar ratio of V_2O_5 60 mole % or above. This may be due to increase in electron hopping rate. They also reported electrical properties for PbO-V_2O_5 glass system with PbO ranging from 10 to 40 mole % [37]. The results have been fitted to various models for AC and DC conduction and the validity of these models in the vanadate glass systems of different compositions has been discussed. Mandal *et al.* carried out thermoelectric power measurements on (100-x) Bo: xV_2O_5 glass systems for values of x in the 90 mole % - 70 mole % range [38]. These glasses are found to be n-type with a value of See beck co-efficient which is temperature independent. Recently, Hayakawa *et al.* [39] carried out IR and NMR studies on $xPbO$-$(100-x)V_2O_5$ glass systems (x = 30, 40 and 50 mole %). Their results indicate that these glass systems consist of VO4 octahedra and VO5 trigonal bipyramids. There are several studies on MO-V_2O_5 (when MO is divalent metal oxide other than PbO [40 and other references there in] or $YnOm$ - V_2O_5 (when YnOm is a glass forming agent like B_2O_3 or Bi_2O_3 [41, 42]. Dimitriev *et al.* have studied glass formation in various $MnOm$-V_2O_5 binary vanadate glass systems (MnOm being a metal oxide like Li2O or ZnO or ZrO2) [43]. They reported composition (mole %) up to which the metal oxide formed glass with V_2O_5. However, of all the binary vanadate glass systems containing divalent metal oxide, the PbO -V_2O_5 glass system has been most widely studied. However, there are a few studies on ternary and quaternary vanadate glass systems. For example, Sumita *et al.* studied V_2O_5 - BaO - ZnO glass system in which V_2O_5 is varied from 30 mole % - 70 mole % [44, 45]. Kawamoto extended these studies to V_2O_5 - BaO - K_2O - ZnO [46] and a pseudo binary system V_2O_5 – (BaO)½ - (ZnO)½ [47].

Even though these studies are interesting because of the complexity of the glass system they do not offer much structural information which can be correlated to the physical properties.

Sen and Ghosh [48] studied electrical conduction of calcium vanadate glasses in 80-500 K. It has been found that the multi-phonon assisted hopping model of small polarons in the non-adiabatic regime can be used to interpret the conductivity data in these systems. Bhattacharya and Ghosh [49] studied electrical properties of a semi-conducting silver vanadate glasses in the temperature range 93 to 423 K. These glasses exhibit much high conductivities when compared to traditional vanadate glasses. In these glasses also multi-phonon assisted hopping model of small polarons in the non-adiabatic regime was found to be consistent with the temperature dependence of conductivity in the entire temperature range. High field and high pressure behavior of $xCuO: (45-x)PbO: 55V_2O_5$ (x = 0 to 20) glasses was investigated by Vaidhyanathan *et al* [50]. The glasses with x = 10 were found to be exhibiting a current controlled negative resistance behavior associated with a memory type transition. These glasses with x = 15 were found to be showing a threshold type behavior.

Harshvadan *et al.* [51], investigated switching behavior of potassium boro-vanadate glasses containing 5 mole % of Fe_2O_3. It was observed that the switching temperature decreases as V_2O_5 concentration increases. Detailed studies of electrical conductivity [52], infrared and optical absorption [53] and switching characteristics [54] of $B_2O_3-V_2O_5$ glasses doped with Fe_2O_3 were studied by El. Mansour *et al.*, Acoustical properties of lead vanadate glasses were studied in the temperature range of 80 to 300K by Paul *et al.* [55]. The temperature dependence of acoustic attenuation could be interpreted in terms of thermally activated relaxation processes. The variation of velocity was explained by the combination effect of relaxation an-harmonicity, and frozen in fluctuations causing in-homogeneity of elasticity in these glass systems. Amnerkar*et al.* [56] studied four different compositions of $B_2O_3-V_2O_5$ glasses at 9.586 GHz for AC conductivity dielectric constant and loss factor. These $B_2O_3-V_2O_5$ glasses were supposed to have potential application as dielectric resonators. Several structural investigations have also been carried out on vanadate glasses more recently. Structure of alkaline earth vanadate glasses were reported from x-ray as well as neutron diffraction techniques to get an idea of change in V – O coordination number in V_2O_5 [36]. Structure of vanadium tellurite glasses were reported by several workers. They employed Raman and XAFS spectroscopies to study coordination changes around Te^{4+} and V^{5+} ions [57]. X-ray photo-electron spectroscopic studies conducted on lead vanadate glasses yielded structural information regarding V – O – V, V– O – Pb and Pb – O – Pb bonds with oxygen as bridging atom [58]. There has been an increased interest in the study of electrical conductivity of devitrified vanadate glasses. Nishada studied electrical conductivity in $25K_2O$ - $65V_2O_5$ - $10Fe_2O_3$ system in glass as well as devitrified state. It has been observed that conductivity jumps from the order of 10^{-8} in the glass to 10^4 in the devitrified samples [59]. A significant increase in electrical conductivity from the order of 10^{-7} to 10^{-2} S/cm was found in Fe_2O_3 doped B_2O_3–V_2O_5 glass system when it was devitrified at the crystallization temperature [59]. The increase in conductivity in the devitrified state was attributed to increase in structural relaxation in crystalline solid when compared to their amorphous state.

Bismuth lead vanadate was another system studied in crystalline form by x-ray and neutron diffraction techniques. This material which is supposed to be useful for 2nd harmonic generation is found to be undergoing a structural phase transition from a non-centro-symmetric phase to a centro-symmetric high temperature phase [60]. La_2O_3 and Fe_2O_3 are added to PbO-V_2O_5 chosen at 1:1, 2:1 and 1:2 molar ratios [61]. These studies using FTIR and laser Raman indicated crystallization of $Pb_2V_2O_7$ inside the glass matrix with increased La_2O_3 substitution. PbO-V_2O_5 glass system has been extensively studied and since the crystal structures are known for PbV_2O_6 or $Pb_2V_2O_7$ *etc.*, this is an ideal system to study the effect of doping or substitution on the physical properties. Any change in the physical properties can be ascribed to the possible changes in the glass structure especially when doping or substitution is carried out at specific compositions for which structural information is available. Ramesh [62] studied ZnO, CuO and TiO_2, substituted by equal molar ratios for PbO in PbO: V_2O_5 glass system chosen at the eutectic composition *i.e.*, 1:1 molar ratio. Some very interesting results like conversion from n-type semi-conduction to p-type semi-conduction in ZnO and TiO_2 substituted glass systems as temperature was increased at a given composition were obtained whereas Cu substituted samples were found to remain n type at all temperatures for a given composition like the un-substituted $50PbO$: $50V_2O_5$ glass system. Recently Tejeswararao *et al.* [63] reported the DSC studies on x Bi_2O_3–$(50-x)PbO$–$50V_2O_5$ glass systems indicated that Bi_2O_3 is substituting for PbO without drastically affecting the microstructure of the unsubstituted lead metavanadate glass network up to x = 15 mol %. As annealing temperature is increased, the increase in the dc conductivity of the present glass systems along with a change in the activation energies can be attributed to a change in the microstructure. The observation of exchange broadening of the ESR spectra recorded at 300 K for the glass samples which were annealed at 150°C is in accordance with the observation of a reduction in the activation energy and an increase in the conductivity of the Bi_2O_3 substituted glass samples when compared to the values obtained for unsubstituted 50 V_2O_5:$50PbO$ glass system. The different oxidation states of Bi are supposed to be acting as electron traps in reducing the electrical conductivity and increasing the activation energy for dc conductivity of x Bi_2O_3–$(50-x)PbO$–50 V_2O_5 (x = 5%, 10%, 15%) glass systems annealed at 225°C. With an increase in the measurement temperature these traps seem to be removed giving rise to an increase in conductivity. Even when the annealing temperature is increased to 225°C, Bi_2O_3 seems to be substituting for PbO in the glass network as can be seen from the x-ray spectra of the devitrified samples. The Hruby parameter indicates that the glass system with 15 mole% of Bi_2O_3 substitution has maximum glass forming ability.

Electrical conduction in the vanadate or other transition metal ion containing semi-conducting glass systems is due to the hopping of electrons between transition metal ion sites of different valence states (V4+, V5+, *etc.* ions) [64 - 69]. The general mechanism of electrical conduction in these glass systems involves formation of small polarons. From the D.C. electrical conductivity measurements, it is possible to evaluate the activation energy, nature of hopping mechanism and other physical parameters pertaining to polaron formation. Different theoretical models [65, 66, 70, 71] study electrical conduction in amorphous semiconductors. Majority of them deal with the mechanism of phonon and polaron interaction as well as explanation of temperature dependence of conductivity. According to model given by Mott conduction takes place between localized states due to small polaron hopping helped by lattice phonons.

The DC conductivity is given by:

$$\sigma = v_0 \left[\frac{e^2 C(1-C)}{k_B T} \right] \, Exp(-2\alpha R) \, Exp\left(-\frac{W}{k_B T} \right) \qquad (9.9)$$

for hopping between nearest neighbors at temperature $T > \theta_D/2$

where v_0 is the longitudinal optical phonon frequency,

R the average site separation,

α^{-1} the spatial decay parameter for s-like wave function assumed to describe the localized state at each site,

C the ratio of the concentration of TMI in low valence state to the total TMI concentration. (*i.e.* the fraction of sites occupied by the polaron)

W the activation energy for DC conduction and

θ_D is the Debye temperature given by $h v_0 = k_B \theta_D$.

Assuming that a strong electron-phonon interaction exists, the activation energy W is given by:

$W = W_H + (W_D/2)$ for $T > \theta_D/2$ where

W_H is the polaron hopping energy given by $W_H \approx W_p/2$,

Wp is the polaron binding energy, and W_D is disorder energy arising from the variation of the local arrangement of ions. The activation energy $W \cong W_D$ for $T < \theta_D/4$.

When the tunneling term exp $(-2\alpha R) \approx 1$, the adiabatic limit is obtained.

At lower temperature $(T < \theta_D/4)$ where the polaron binding energy is considerably small, the disorder energy W_D plays a dominant role in the conduction mechanism. In such a case Mott and Austin have proposed that hopping may occur preferentially beyond nearest neighbours (with same W_D) which is called variable range hopping. The temperature dependence of DC conductivity can be expressed as:

$$\sigma = \sigma_0 \, \mathrm{Exp}\left[-\left(\frac{T_o}{T}\right)^{\!1/4} \right] \tag{9.10}$$

where σ_o and T_o are constants and,

$$T_0 = \frac{19.4 \; \alpha^3}{k_B N(E_F)} \tag{9.11}$$

where $N(E_F)$ is the density of states at the Fermi level.

The value of polaron radius r_p is obtained from the following formula given by Bogomolov *et al.* [72], for the case of a non-dispersive system of frequency v_0

$$r_p = \left(\frac{\pi}{6}\right)^{\!1/3} R/2 \tag{9.12}$$

Schnakenberg has considered a more general polaron hopping model [73] in which optical multi-phonon processes determine the DC conductivity at high temperatures, while at low temperatures charge transport is described by acoustical one phonon assisted hopping process. The expression for the DC conductivity is given by:

$$\sigma \cong \frac{1}{T}\left[\text{Sinh}\left(\frac{h\nu_o}{k_BT}\right)\right]^{\frac{1}{2}} \text{Exp}\left[-\left(\frac{4W_H}{h\nu_o}\right)\tanh\left(\frac{h\nu_o}{4k_BT}\right)\right] \times \text{Exp}\left[-\left(W_D/k_BT\right)\right] \qquad (9.13)$$

This equation predicts a temperature dependent hopping energy which decreases with the decrease of temperature. While fitting the conductivity data to Mott's model, it has been found in literature [42] that for a given series of glass compositions a plot of logarithm of conductivity at a given experimental temperature as a function of the activation energy at that temperature is a straight line if hopping is adiabatic. If the straight-line pattern is not obtained, the hopping is supposed to be non-adiabatic [37, 74, 75].

Thermoelectric power is another transport property closely related to σ (the D.C. conductivity) and gives considerable information regarding transport mechanisms in crystalline or amorphous semiconductors. The thermopower or See beck coefficient, denoted by S, is a constant of proportionality between the voltages (in the absence of a current) developed between the two parallel faces of the sample because of the temperature gradient applied between those two faces. The thermoelectric power measurements of the transition metal oxide glasses [76] are of great interest because of the information they yield on the nature of charge carriers, polaron formation, the disorder energy due to random fields *etc.*

In a wide band semi-conductor, the thermoelectric power is given by [66].

$$S = \left(\frac{K_B}{e}\right)\left\{\ln\left(\frac{n}{P}\right) + \alpha'\right\} \qquad (9.14)$$

where

k_B is the Boltzmann's constant

e is the charge on the carrier (electrons)

n is the number of electrons per unit volume

$P = 2 (N_0 - n) k'$ where

N_o is the number of sites per unit volume available for the electron

k' is the ratio of the number of acceptors to donors (k'< 1)

and α' is the function of KBT and indicates the kinetic energy of a carrier.

Another mathematical expression for the thermoelectric power involving impurity conduction in semi-conductors or glasses containing transition metal ions with mixed valance states has been given by Heikes [77] as follows:

$$S = \left(\frac{K_B}{e}\right)\left\{\ln\left(\frac{C}{(1-C)}\right) + \alpha'\right\} \tag{9.15}$$

where C the fraction of ions in reduced valance states and the other terms with usual meaning. According to Heikes [77], $\alpha' = \frac{\Delta S'}{K_B}$, where $\Delta S'$ is the change in entropy of an ion due to the presence of a charge carrier. Austin and Mott [65] have shown that $\alpha' = \sum \Delta \omega_o / \omega_o$ where $\Delta \omega_o$ is the change in the characteristic phonon frequency ω_o of a given ion because of perturbation caused due to presence of a charge carrier with summation extending over all neighbouring ions.

Austin and Mott [66] suggested that α' ≥ 2 for large polaron, while, as suggested by Mott [65], small polaron formation is indicated by α' < 1. Thus, the magnitude of α' can be used to know the formation of small or large polaron that takes place in materials. As per Austin and Mott [66] α' in Eq. (15) is given by

$$\alpha' = \frac{(1-\theta)W_H}{(1+\theta)K_B T} \tag{9.16}$$

where W_H is the polaron hopping energy and θ is a constant correlated to the amount of disorder in the system. θ = 1 corresponds to zero disorder energy and any deviation from this value corresponds to the amount of disorder in the glass samples. For conduction in a material having an impurity bandwidth $> K_B T$, the following formula is applicable [66, 78].

$$S = \left(\pi^2 K_B^2 \frac{T}{3e}\right) \left[d(\ln \sigma)/dE\right]_{E=E_F} \tag{9.17}$$

Using $\sigma(E) = \sigma_o(E) \exp\left(\dfrac{-W_D}{K_B T}\right)$ as the temperature dependent conductivity one can readily get from Eq. (17) the following expression for S.

$$S = \frac{\pi^2 K_B}{3e}\left(\frac{d \ln \sigma_o}{dE} - \frac{dW_D}{dE}\right) \tag{9.18}$$

which is valid for $k_B T \ll W_D$ where disorder energy is given by W_D .

It is seen that the energy required to produce equal number of carriers at low temperature is called disorder energy. As per Emin [79], thermoelectric power due to charge 'e' that moves in molecular crystal lattice is given by

$$S = -\left(\frac{K_B}{e}\right)\left(\frac{W_D}{K_B T}\right) = -\frac{W_D}{eT} \tag{9.19}$$

Similar equation for temperature dependent energy gap is given by Mott [13]

$$S = \frac{K_B}{e}\left(\frac{E_o}{K_B T} - \frac{\gamma}{K_B} + 1\right)$$

where E_o is the energy gap at 0 K. γ is the coefficient of temperature dependence of the energy gap. With increase of temperature ZnO and TiO_2 substituted samples change from n-type to p-type. The CuO substituted system remains n-type like the unsubstituted system even at high temperatures ($\approx$ 450 K). The amount of disorder was same for all systems when Heike's formula is applied at room temperature. Estimation of W_D with Emin's formula in ZnO, TiO_2 substituted samples, large value $\sim$ 0.6 eV was obtained. The present results indicate that the Emin's formula cannot be directly used to estimate the disorder energy in ZnO and TiO_2 substituted lead vanadates. Change in V^{4+}/V^{5+} ratio with change in temperature might change the sign of the thermoelectric power S, in these systems. To establish these alternative mechanisms further studies on these systems are required.

CONCLUSIONS

Perfect devitrification can be obtained for all glass samples by using melt quenching technique on a stainless steel and graphite blocks. DSC recording of vitrified and XRD of devitrified samples indicate possible formation of various meta-stable and stable phases. DC conductivity studies indicated that conduction is due to small polaron hopping. Present study suggest application of Mott's variable range hopping model at room temperature, Mott's phonon assisted hopping process in non-adiabatic regime at high temperatures and Schnackenberg's generalized polaron hopping model for conduction mechanism in these glasses. As per thermo emf measurements all glass samples are n-type at normal temperature and with increase in temperature CuO, Ag_2O and ZnO substituted systems remain to be n type similar to previous mentioned undoped systems even up to 500° K temperature. In sharp contrast the CdO, TiO_2 and TeO_2 substituted samples change from n-type to p-type as the temperature is increased. When Heike's formula is applied to all the systems at room temperature the amount of disorder in all these systems was found to be nearly same. When Emin's formula is used for the estimation of W_D, the activation energy due to disorder, in such kind of samples, unusually large values of W_D (~ 0.22 eV) are obtained. The temperature dependent change of sign of the thermoelectric power S, in these systems may arise due to change in V^{4+}/V^{5+} ratio with the change of temperature or due to the onset of band type of conduction as in MnO or other extrinsic compensated semi-conductors. To establish these alternative mechanisms further studies on these systems are required.

CONSENT FOR PUBLICATION

Not applicable.

CONFLICT OF INTEREST

The authors confirm that this chapter content has no conflict of interest.

ACKNOWLEDGEMENTS

The authors express thankfulness to Dr. P. Sreeramulu, Assistant Professor (English), GITAM, Bangalore for providing English language editing services to this manuscript.

REFERENCES

[1]   S.R. Ellioit, *Physics of Amorphous Materials.* 1st ed. Longman: London, 1984.

[2]   A.E. Owen, *Electronic and Structural Properties of Amorphous Semi-Conductors.*, P.G. Lecomber, J. Mort, Eds., Academic press: London, 1973.

[3]   P.W. Anderson, "Absence of Diffusion in Certain Random Lattices", *Phys. Rev.,* p. 109, 1958.
[http://dx.doi.org/10.1103/PhysRev.109.1492]

[4]   N.F. Mott, "The electrical properties of liquid mercury", *Philos. Mag.,* vol. 13, p. 989, 1966.
[http://dx.doi.org/10.1080/14786436608213149]

[5]   N.F. Mott, "Electrons in disordered structures", *Adv. Phys.,* vol. 16, p. 49, 1967.
[http://dx.doi.org/10.1080/00018736700101265]

[6]   H.L. Frisch, and S.P. Lloyd, "Electron Levels in a One-Dimensional Random Lattice", *Phys. Rev.,* vol. 120, p. 1175, 1960.
[http://dx.doi.org/10.1103/PhysRev.120.1175]

[7]   I.M. Lifshitz, "The energy spectrum of disordered systems", *Adv. Phys.,* vol. 13, p. 483, 1964.
[http://dx.doi.org/10.1080/00018736400101061]

[8]   M.H. Cohen, H. Fritzche, and S.R. Ovshinsky, "Simple Band Model for Amorphous Semiconducting Alloys", *Phys. Rev. Lett.,* vol. 22, p. 1065, 1969.
[http://dx.doi.org/10.1103/PhysRevLett.22.1065]

[9]   E.A. Davis, and N.F. Mott, "Conduction in non-crystalline systems V. Conductivity, optical absorption and photoconductivity in amorphous semiconductors", *Philos. Mag.,* vol. 22, p. 903, 1970.
[http://dx.doi.org/10.1080/14786437008221061]

[10]   N.F. Mott, "Evidence for a pseudo gap in liquid mercury", *Philos. Mag.,* vol. 26, p. 505, 1972.
[http://dx.doi.org/10.1080/14786437208230101]

[11]   J.M. Marshal, and A.E. Owen, "Drift mobility studies in vitreous arsenic triselenide", *Philos. Mag.,* vol. 24, p. 1281, 1971.
[http://dx.doi.org/10.1080/14786437108217413]

[12]   R.A. Street, and N.F. Mott, "States in the Gap in Glassy Semiconductors", *Phys. Rev. Lett.,* vol. 35, p. 1293, 1975.
[http://dx.doi.org/10.1103/PhysRevLett.35.1293]

[13]   N.F. Mott, E.A. Davis, and R.A. Street, "States in the gap and recombination in amorphous semiconductors", *Philos. Mag.,* vol. 32, p. 961, 1975.
[http://dx.doi.org/10.1080/14786437508221667]

[14]   N.F. Mott, and E.A. Davis, *Electronic Processes in Non - Crystalline Materials.* 2nd ed. Clarendon Press: Oxford, 1979.

[15]   S.R. Ellioit, *Physics of Amorphous Materials.* 1st ed. Longman: London, 1984.

[16]   R. Zallen,

[17]   S.R. Ovshinsky, "Reversible Electrical Switching Phenomena in Disordered Structures", *Phys. Rev. Lett.,* vol. 21, p. 1450, 1968.
[http://dx.doi.org/10.1103/PhysRevLett.21.1450]

[18]   C.F. Drake, I.F. Scanlon, and A. Engel, "Electrical Switching Phenomena in Transition Metal Glasses under the Influence of High Electric Fields", *Phys. Status Solidi,* vol. 32, p. 193, 1969.
[http://dx.doi.org/10.1002/pssb.19690320121]

[19]   R.A. Montani, M. Levy, and J.L. Souquet, An electrothermal model for high-field conduction and

switching phenomena in TeO2- V_2O_5 glasses., *J. Non-Cryst. Solids,* vol. 149, p. 249, 1992.
[http://dx.doi.org/10.1016/0022-3093(92)90073-S]

[20]   A.E. Owen, "Semiconducting glasses part 1: Glass as an electronic conductor", *Contemp. Phys.,* vol. 11, p. 227, 1970.
[http://dx.doi.org/10.1080/00107517008202179]

[21]   E.P. Denton, H. Rawson, and J.E. Stanworth, "Vanadate Glasses", *Nature,* vol. 173, p. 1030, 1954.
[http://dx.doi.org/10.1038/1731030b0]

[22]   P.L. Baynton, H. Rawson, and J.E. Stanworth, "Semiconducting Properties of Some Vanadate Glasses", *Trans. Electrochem. Soc.,* vol. 104, p. 237, 1956.
[http://dx.doi.org/10.1149/1.2428544]

[23]   M. Amadori, "Attl. r. 1st", *Veneto,* vol. 76, p. 419, 1917.

[24]   P. Gregeire, D. Josso, P. Garnier, and E.T.G. Calvarin, "Nouveandiagramme de phase pour ie system 91-x) V_2O_5:xPbO diesel domaine x <2/3", *J. Solid State Chem.,* vol. 64, p. 225, 1986.

[25]   E. Baiocchi, M. Bettinelli, and A. Motenero, "New phases in equimolar PbO: V_2O_5 system", *J. Solid State Chem.,* vol. 43, p. 63, 1982.
[http://dx.doi.org/10.1016/0022-4596(82)90215-8]

[26]   G. Calestani, A. Montenero, F. Pigoli, and M. Bettinelli, "Glassy and crystalline phases in the PbO: V_2O_5 system", *J. Solid State Chem.,* vol. 59, p. 357, 1985.
[http://dx.doi.org/10.1016/0022-4596(85)90303-2]

[27]   F. Momo, A. Sotgin, and E. Baicchi, "ESR study at the equimolarPbO: V_2O_5 system", *J. Mater. Sci.,* vol. 17, p. 3221, 1982.
[http://dx.doi.org/10.1007/BF01203486]

[28]   E. Agostinelli, P. Filaci, and D. Fiorani, "Magnetic properties of vitreous and crystalline Pb V2O6", *J. Non-Cryst. Solids,* vol. 84, p. 329, 1986.
[http://dx.doi.org/10.1016/0022-3093(86)90794-5]

[29]   B.D. Jordan, and C. Calvo, "Crystal structure of lead meta vanadate PbV2O6", *Can. J. Chem.,* vol. 52, p. 2701, 1974.
[http://dx.doi.org/10.1139/v74-393]

[30]   G. Calestani, G.D. Andreetti, and A. Montenero, "Structure of lead meta vanadates the monoclinic PbV2O6 (II) modification", *Acta Crystallogr.,* vol. C41, p. 177, 1985.

[31]   G. Calestani, G.D. Andreetti, A. Montenero, and J. Rebizant, "Structure of lead meta vanadates the orthorhombic PbV2O6 (III) modification", *Acta Crystallogr.,* vol. 41, p. 179, 1985.

[32]   V. Fares, M. Magini, and A. Montenero, "X-ray diffraction investigation of the structure of lead metavanadate PbV2O6 glass", *J. Non-Cryst. Solids,* vol. 99, p. 404, 1988.
[http://dx.doi.org/10.1016/0022-3093(88)90446-2]

[33]   A.C. Wright, C.A. Yarker, P.A.V. Johnson, and R.N. Sindair, "A Neutron diffraction investigation of the structure of phosphorous barium and lead vanadates glasses", *J. Non-Cryst. Solids,* vol. 76, p. 333, 1985.
[http://dx.doi.org/10.1016/0022-3093(85)90009-2]

[34]   Y. Dimitriev, V. Dimitrov, M. Arnandov, and D. Topalov, "IR-spectral study of vanadate vitreous systems", *J. Non-Cryst. Solids,* vol. 57, p. 147, 1983.
[http://dx.doi.org/10.1016/0022-3093(83)90417-9]

[35]   V. Dimitrov, and Y. Dimitriev, Structure of glasses in PbO-V_2O_5 system., *J. Non-Cryst. Solids,* vol. 122, p. 133, 1990.
[http://dx.doi.org/10.1016/0022-3093(90)91058-Y]

[36]   S. Mandal, and A. Ghosh, "Structure and physical properties of glassy lead vanadates", *Phys. Rev. B Condens. Matter,* vol. 48, no. 13, pp. 9388-9393, 1993.

[http://dx.doi.org/10.1103/PhysRevB.48.9388] [PMID: 10007177]

[37]　S. Mandal, and A. Ghosh, "Electrical properties of lead vanadate glasses", *Phys. Rev. B Condens. Matter,* vol. 49, no. 5, pp. 3131-3135, 1994.
[http://dx.doi.org/10.1103/PhysRevB.49.3131] [PMID: 10011170]

[38]　S. Mandal, D. Benerjee, R.N. Bhattacharya, and A. Ghosh, "Thermoelectric power of unconventional lead vanadate glass, J. Phys", *Condems. Matter,* vol. 8, p. 2865, 1996.
[http://dx.doi.org/10.1088/0953-8984/8/16/016]

[39]　S. Hayakawa, T. Yoko, and S. IR Sakka, *J. Non-Cryst. Solids,* vol. 183, p. 73, 1995.
[http://dx.doi.org/10.1016/0022-3093(94)00652-0]

[40]　V. Dimitrov, "Strucural changes in vitreous vanadate systems", *J. Non - Cryst. Solids,* vol. 183, pp. 192-193, 1995.

[41]　H. Mkami, and El, "Dc and ac Conductivities of (V_2O_5) x(B_2O_3)1−x oxide glasses Phys. Chem. Glasses", *Phys. Chem. Glasses,* vol. 38, p. 137, 1997.

[42]　A. Ghosh, and B.K. Chaudari, "DC conductivity of V_2O_5-Bi2O3 glasses", *J. Non-Cryst. Solids,* vol. 83, p. 151, 1986.
[http://dx.doi.org/10.1016/0022-3093(86)90065-7]

[43]　Y. Dimitriev, I. Ivanova, and E. Gatev, *J. Non-Cryst. Solids,* vol. 45, p. 293, 1987.

[44]　S. Sumita, K. Horachi, M. Toyama, and S. Truchihashi, *J. Ceram. Soc. Jpn.,* vol. 82, p. 57, 1974.

[45]　S. Sumita, K. Horachi, M. Toyama, and S. Truchihashi, *Proc. Xth Int. Congr. Glass,* vol. 13, 1974p. 36

[46]　Y. Kawamoto, M. Fukuzuka, Y. Ohta, and M. Imai, "Electronic conduction and glass structure in V_2O_5 –BaO-K_2O-ZnO glasses", *Phys. Chem., glass,* vol. 20, p. 54, 1979.

[47]　Y. Kawamoto, J. Tanida, H. Hamada, and H. Kiriyama, "Vanadium bronze-vanadate glasses in the system $V_2O_5 – V_2O_4$Ba1/2-Zn1/4 and other magnetic and electrical properties", *J. Non-Cryst. Solids,* vol. 38, p. 301, 1980.
[http://dx.doi.org/10.1016/0022-3093(80)90435-4]

[48]　S. Sen, and A. Ghosh, "Hopping conduction in calcium vanadate semiconducting glasses", *J. Phys. Condens. Matter,* vol. 11, p. 8061, 1999.
[http://dx.doi.org/10.1088/0953-8984/11/41/309]

[49]　S. Battacharya, and A. Ghosh, "Polaron transport in semiconducting silver vanadate glasses", *Phys. Rev.,* vol. 66, 2002.132203
[http://dx.doi.org/10.1103/PhysRevB.66.132203]

[50]　B. Vaidhyanathan, S. Asokan, and K.J. Rao, "Effect of CuO addition on the high field and high-pressure behavior of microwave prepared lead vanadate glasses", *J. Mater. Res.,* vol. 15, p. 518, 2002.
[http://dx.doi.org/10.1557/JMR.2000.0077]

[51]　R. Harshvadan, "Panchal and Dinesh K. Kanchan, Structural, physical, and electrical properties of boro-vanadate-iron glasses doped with K_2O alkali", *Turk. J. Phys.,* vol. 23, p. 969, 1999.

[52]　I.A. Gohan, Y.M. Moustafa, A.A. Megahed, and El. Mansour, "Electrical properties of semi-conducting barium-vanadate glasses containing Iron oxide", *Phys. Chem. Glasses,* vol. 39, pp. 56-60, 1998.

[53]　Y.M. Moustafa, I.A. Gohan, A.A. Megahed, and El. Mansour, "Spectroscopic studies of semi conducting barium-vanadate glasses", *Phys. Chem. Glasses,* vol. 38, p. 92, 1997.

[54]　I.A. Gohan, Y.M. Moustafa, A.A. Megahed, and El. Mansour, "Switching characteristics of conducting barium-vanadate glasses containing Iron oxide", *Phys. Chem. Glasses,* vol. 38, p. 37, 1997.

[55]　A. Paul, A. Malti, and C. Basu, "Acoustic properties of lead-vanadate glasses", *J. Appl. Phys.,* vol. 86, p. 3598, 1999.

[http://dx.doi.org/10.1063/1.371265]

[56]　R.H. Amnerkar, C.S. Adgoankar, S.S. Yawale, and S.P. Yawale, "Microwave properties of vanadium borate glasses", *Bull. Mater. Sci.,* vol. 25, p. 431, 2002.
[http://dx.doi.org/10.1007/BF02708022]

[57]　D. Bersani, G. Antonioli, and P.P. Lottici, "Coordination changes in telluro-vanadate glasses containing ZnO or CdO", *J. Non-Cryst. Solids,* vol. 234, p. 293, 1998.
[http://dx.doi.org/10.1016/S0022-3093(98)00494-3]

[58]　A. Mekki, "XPS study of lead vanadate glasses", *Arab. J. Sci. Eng.,* vol. 28, p. 73, 2003.

[59]　K. Fukuda, A. Ikeda, and T. Nishida, "A significant improvement of the electrical conductivity of semiconducting vanadate glasses caused by heat treatment", *Diffus. Defect Data Solid State Data Pt. B Solid State Phenom.,* vol. 90, p. 215, 2003.
[http://dx.doi.org/10.4028/www.scientific.net/SSP.90-91.215]

[60]　I.R. Evans, J.S.O. Evans, and J.A.K. Howard, "Variable temperature structural study of bismuth lead vanadate, BiPb2VO6", *J. Mater. Chem.,* vol. 12, p. 2648, 2002.
[http://dx.doi.org/10.1039/b202648a]

[61]　S. Music, M. Gotic, and S. Popovic, "Structural properties of lead vanadate glasses containing La3+ or Fe3+ ions", *J. Mater. Sci.,* vol. 29, p. 1227, 1994.
[http://dx.doi.org/10.1007/BF00975069]

[62]　K.V. Ramesh, D.L. Sastry, and Y. Purushotham, "Temperature-dependent thermoelectric power of Cuo, ZnO and TiO_2 substituted for PbO in eutectic lead vanadate glass system", *Int. J. Mod. Phys. B,* vol. 18, p. 3327, 2004.
[http://dx.doi.org/10.1142/S0217979204026482]

[63]　Tejeswara rao P, Ramesh K. V. and Sastry D. L, "ESR and DC electrical properties of $Bi2O3$-PbO-V_2O_5 glass system", *Phys. Chem. Glasses: Eur. J. Glass Sci. Technol. B,* vol. 57, p. 279, 2016.

[64]　M. Sayer, and A. Mansingh, "Transport properties of semiconducting phosphate glasses", *Phys. Rev.,* vol. 6, p. 5629, 1972.

[65]　N.F. Mott, "Conduction in glasses containing transition metal ions", *J. Non-Cryst. Solids,* vol. 1, p. 1, 1968.
[http://dx.doi.org/10.1016/0022-3093(68)90002-1]

[66]　I.G. Austin, and N.F. Mott, "Polarons in crystalline and non-crystalline materials", *Adv. Phys.,* vol. 18, p. 41, 1969.
[http://dx.doi.org/10.1080/00018736900101267]

[67]　C.H. Chung, and J.D. Mackenzie, "Electrical properties of binary semiconducting oxide glasses containing 55 mole % V_2O_5", *J. Non-Cryst. Solids,* vol. 42, p. 357, 1980.
[http://dx.doi.org/10.1016/0022-3093(80)90036-8]

[68]　A.E. Owen, "The electrical properties of glasses", *J. Non-Cryst. Solids,* vol. 25, p. 370, 1977.
[http://dx.doi.org/10.1016/0022-3093(77)90099-0]

[69]　A. Ghosh, "Electrical transport properties of molybdenum tellurite glassy semiconductors", *Philos. Mag.,* vol. 61, p. 87, 1990.
[http://dx.doi.org/10.1080/13642819008208653]

[70]　A.R. Long, "Frequency-dependent loss in amorphous semiconductors", *Adv. Phys.,* vol. 31, p. 553, 1982.
[http://dx.doi.org/10.1080/00018738200101418]

[71]　S.R. Elliott, "A.C. conduction in amorphous chalcogenide and pnictide semiconductors", *Adv. Phys.,* vol. 36, p. 135, 1987.
[http://dx.doi.org/10.1080/00018738700101971]

[72]　V.N. Bogomolov, E.K. Kudinov, and T.A. Firsov, "Polaron nature of the current carriers in rutile TiO2, Sov", *Phys. Solid State,* vol. 9, p. 2502, 1968.

[73]　J. Schnakenderg, *Polaronic Impurity Hopping Conduction, Phys.* Status Solids, 1968, p. 623.

[74]　S. Mandal, and A. Ghosh, "Electrical conduction in lead - iron glasses", *J. Phys. Condens. Matter,* vol. 8, p. 829, 1996.
　　　[http://dx.doi.org/10.1088/0953-8984/8/7/008]

[75]　A. Ghosh, "Transport properties of vanadium germanate glassy semiconductors", *Phys. Rev. B Condens. Matter,* vol. 42, no. 9, pp. 5665-5676, 1990.
　　　[http://dx.doi.org/10.1103/PhysRevB.42.5665] [PMID: 9996152]

[76]　B.W. Flynn, A.E. Owen, and J.M. Robertson, Proc. 7th Conference on Amorphons and Liquid Semi - Conductors,

[77]　R.R. Heikes, *Thermoelectricity,* R.R. Heikes, Ure R.W, Eds., Inter science: New York, 1961, p. 81.

[78]　A. Ghosh, "Temperature dependent thermoelectric power of semiconducting bismuth vanadate glasses", *J. Appl. Phys.,* vol. 65, p. 227, 1989.
　　　[http://dx.doi.org/10.1063/1.342575]

[79]　D. Emin, *Electronic and Structural Properties of Amorphons Semi-Conductors.,* P.G. LeComber, J. Mort, Eds., Academic: New York, 1973, p. 261.

CHAPTER 10

Advanced Ceramics for 3D Printing Applications

B. Venkata Shiva Reddy[1], N. Suresh Kumar[2], K. Chandra Babu Naidu[1,*], Anish Khan[3], Abdullah M. Asiri[3] and B. Kishore[4]

[1] *Dept. of Physics, GITAM Deemed to be University, Bangalore-562163, Karnataka, India*

[2] *Department of Physics, JNTUA, Anantapuramu-515002, A.P, India*

[3] *Chemistry Department, Faculty of Science & Center of Excellence for Advanced Materials Research, King Abdulaziz University, P.O. Box 80203, Jeddah, 21589, Saudi Arabia*

[4] *Dept. of Mechanical Engineering, GITAM Deemed to be University, Bangalore-562163, Karnataka, India*

Abstract: Additive method is employed for 3D printing ceramic materials over a subtractive conventional method with the help of computer-aided design [CAD]. The different feedstocks are used to print ceramic materials such as slurry, powder and bulk solid. Alumina powder is the most common material and excellent oxide used as a catalyst, adsorbent, aerospace material and microelectronics material. With the help of stereolithography technique, glass ceramics and active glass can be printed.

Keywords: Active Glass, Alumina Powder, CAD, Glass Ceramics, Stereolithography.

10.1. INTRODUCTION

Ceramics in a broader concept includes all silicates. The word is derived from the Greek word keramos meaning burnt stuff. Ceramics refer to clay products such as porcelain wares, potteries, refractories, building bricks, insulating materials, *etc.* ceramics is also referred to as inorganic nonmetallic materials processed at high temperatures. The basic raw materials required for the manufacture of ceramics are clay, feldspar, sand or quartz. Ceramic is strong, durability and fire-resistant. The history of ceramics in India dates back to Hindus valley civilization and ceramic materials or articles were prepared by their own traditional methods, during that period, people used the pots for the preparation of food, preservation of grains, *etc.* so, from this, we can say that ceramics or pottery are very ancient

* **Corresponding author Dr. K. Chandra Babu Naidu:** GITAM Deemed To Be University-Bangalore Campus, Bangalore-562163, Karnataka, India; E-mail: chandrababu954@gmail.com

and tool for effective civilization of society. Still in India the pot making is very good cottage industry in Indian village setup. Now a good transition took place in the ceramics, called 3D printing technology, in this technology ideas or dreams become materials with the help of CAD (computer aided designing) *i.e.*, geometrical figures can be processed into desired materials. Compare to traditional method, 3D printing has a high potential to ensure rapid preparation of materials. This technology Prints more complex materials or structures than conventional or traditional materials. This 3D printing is also called as rapid proto typing or additive manufacturing, in this technique the materials are formed by number of layers. In 1980's the first material was prepared at Nagoya municipal industrial research institute. Now varieties of 3D printing materials are printing such as nanomaterials, bio materials, smart materials and membranes and even extended the applications of 3D printing materials in the constructions, heart pumps, aerospace, biomedical, food industry and *etc* [1].

Most commonly used material is alumina powder is excellent ceramic oxide form which is used in catalysts, adsorbents, microelectronics, chemicals and aerospace industry. Stereolithographic machine is used to process glass ceramics and bio active glass and it improves bending strength of the materials [2]. Technology that is preprocessed feedstock before printing generally classified into slurry based, powdered based, and bulk solid based methods that is given in the below Table **10.1**. The viscosity range in slurry-based technology in ceramics/polymer mixtures from low-viscosity (~m Pa· s) inks with a low ceramic loading (up to 30 vol %) to high-viscosity (~Pa· s) pastes with a much greater ceramic loading (up to 60 vol %) [3].

Table 10.1. Technology behind various feedstock forms.

Feed Stock Form	Ceramic 3D Printing Technology Type
Slurry based	Stereolithography [SL] Digital light processing [DLP] Two-photon polymerization [TPP] Inkjet printing [IJP] Direct ink writing [DIW]
Powder based	3-dimensional printing [3DP] Selective LASER sintering [SLS] Selective LASER melting [SLM]
Bulk solid based	Laminated object manufacturing [LOM] Fused deposition modeling [FDM]

The direct ink writing possesses a great application in 3D printing, in the preparations of sensors, electrodes, energy storage, tissue engineering scaffold, and shape morphing system. In this process the most critical one is maintaining the appropriate viscosity and elastic porosity of ink [4]. The digital light processing is also playing an important role in the preparation of meshes for 3D printing materials, that meshes or whiskers are used in the separation of water or oil [with the help of honey comb mesh] and also used to purify oil. This technique is based on the light-curing of light sensitive reaction to produce layer by layer printing and accomplish. From this DLP approach (Table **10.1**), ceramic phosphors also can be prepared for high flux LASER light that is used to fabricate LASER diode which possess high power density due to thermal endurance and thermal stability of luminescence, the material used for this laser is an Al_2O_3/YAG: Ce^{3+} [5, 10, 11]. The working diagram of DLP is as shown in Fig. (**10.1**) [5]. More ecofriendly and naturally occurring polymers are used in the preparation of ceramics one of them is methyl cellulose [MC] which possess low concentration. This methyl cellulose with magnesium aluminate [$MgAl_2O_4$] spinel is used to prepare transparent materials which are generally used in transparent windows and domes, and this spinal is very good refractory. This spinal possesses many bio medical applications such as artificial bones, teeth, hemispherical hip and *etc* [6].

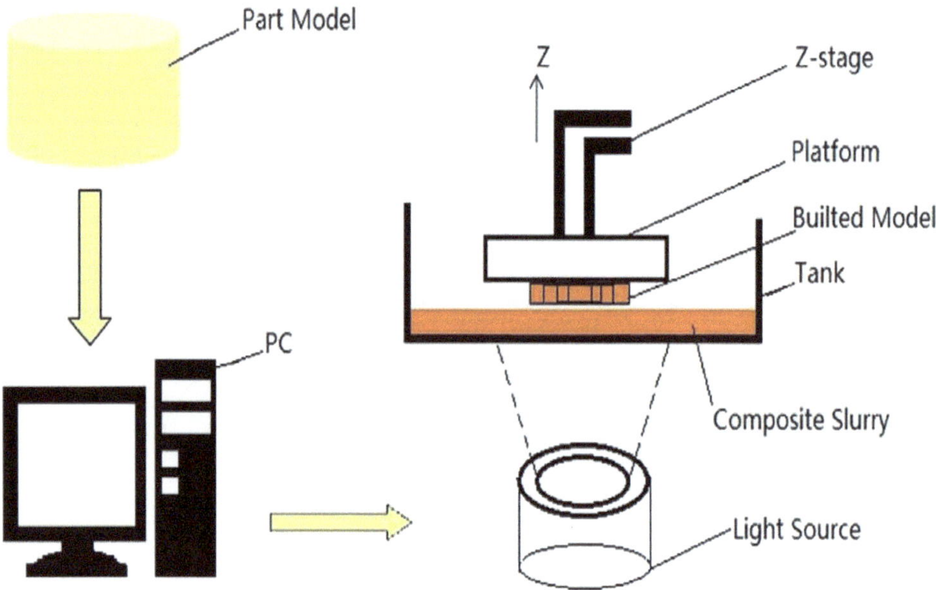

Fig. (10.1). Working diagram of DLP.

In bone making this 3D printing is used (Fig. **10.2**) [7], the materials which are used in bone making are calcium phosphate materials which are similar to our natural bone, such as hydroxyapatite (HAP)$Ca_{10}(PO_4)_6(OH_2)$ and tri calcium phosphate (TCP) $Ca_3(PO_4)_2$. Any way the preparation of ceramic materials has its own shortcomings or drawbacks such as flowability and dispersity of agglomeration which leads to low density and more cracks or flaws in materials. In order to overcome these short comings spray drying is employed and it is rapid and continues process which produces granules with spherical and mono dispersed. The suspension state of materials is important (viscosity, density, and surface tension) in drying process to produce varies materials [7].

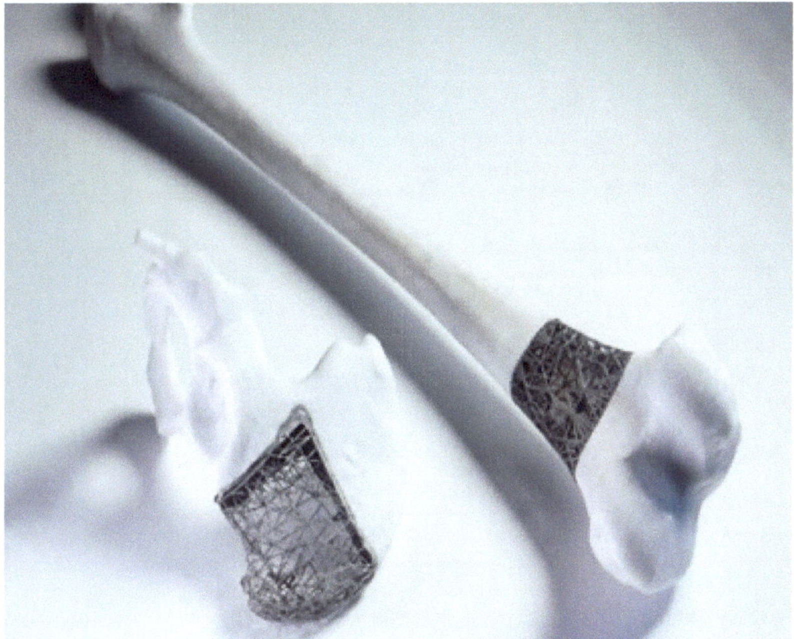

Fig. (10.2). 3D printed bone.

Due to biocompatibility hips are printing in three dimensions which are stronger as they are printed by ceramics, for this printing fused deposition modeling [FDM] will be employed and chemical is polylactic acid [PLA] is used to print [8]. The preparation of nano and micro materials in conventional method is not possible, so we employ stereolithography (Fig. **10.3**) [8] which is matured technique to produce micro and nano scale 3D printings. This method was employed during 1990's [8], scientists prepared Al_2O_3 and SiO_2 and their derivatives. The silica-based materials are also prepared by stereolithography such as SiOC, SiC, *etc.* SiOC ceramic structure is about 200 μm strut size prepared under sintering of 1000°C [9]. For good results and advantages, we can club both DLP and stereolithography techniques because both the techniques depend on the light curing reaction. The most important material can be produced from this method that is zirconia a typical structural ceramic material which is used to prepare hard bearings and also in medical field to prepare bones and teeth [10].

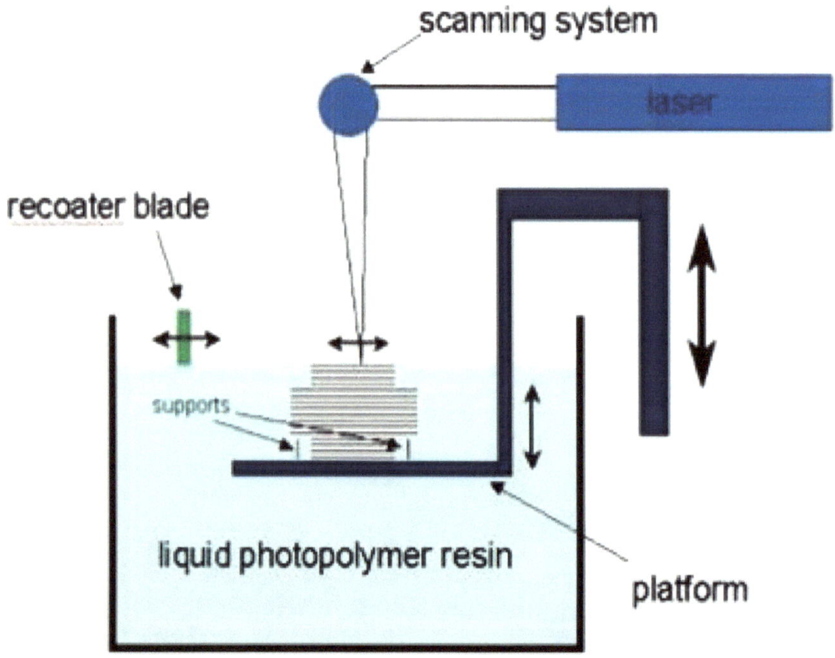

Fig. (10.3). Stereolithography process.

The two-photon polymerization is recent technology which depends on laser technology and used to prepare biomedical 3D material. Here two photons are involved in absorption (2PA), hence the name is given as a two-photon polymerization and molecules excited by only two photons. Each half of the energy that is required to reach the same excited state that would be induced by one-photon absorption. This is possible only when two photons excite same molecules at the same time, this 2PA process is a temporally resolved process therefore high and extremely pulsed lasers are required to get desired spatial resolution. The fabrication of such structures directly from aqueous media avoids any solvent changes, which could impair the accuracy of the structure or even induce cytotoxic effects. Water soluble radical photo initiators (PI), miscible monomers as well as cross linkers are required for this purpose [12]. Selective LASER sintering method is the one of the new methods in 3D printing and laser light is used to sinter powdered particles all together if powder is not sintered then it can be reused for sintering so that we can avoid wastage of powder, its advantages are high resolution and single step printing technology which has been introduced into pharmaceutical sciences. By this technology print-lets (tablets) can be printed [13]. The diameter, thickness and disintegration time of the materials can be found, the diameter and thickness can be determined by digital Vernier calipers, the hardness is also measured by hardness tester (VK 200, Varian Inc, Cary, NC). The disintegration test is conducted by USP disintegration (900 ml water medium, Vankel Varian VK-100, NC, USA) [14].

Selective laser melting [SLM] (Fig. **10.4**) [15] is a powdered based 3D printing technology in which a focused laser beam is driven by galvanometer scanner melts powdered particles. And this powdered particle spread over a substrate to form the layers. This SLM is used to process metallic materials such as titanium, gold, nickel, aluminum, copper, tungsten and related alloys [15]. Laminated object manufacturing [LOM] is the layer-oriented technique which involves formation of layer on the other layer. This layer on the layer behaves as a lamination structure, this process takes place in higher temperature and it requires more than one material [16]. Fused deposition modeling [FDM] is one of the additive manufacturing processes is used to print plastic and plastic composites materials. In this technique plastic is melted and deposited by heated extruder, here material is deposited *via* layer by layer with the help of CAD graphics. These materials generally used in the manufacture of cars, aerospace, energy, and automotive industries [17].

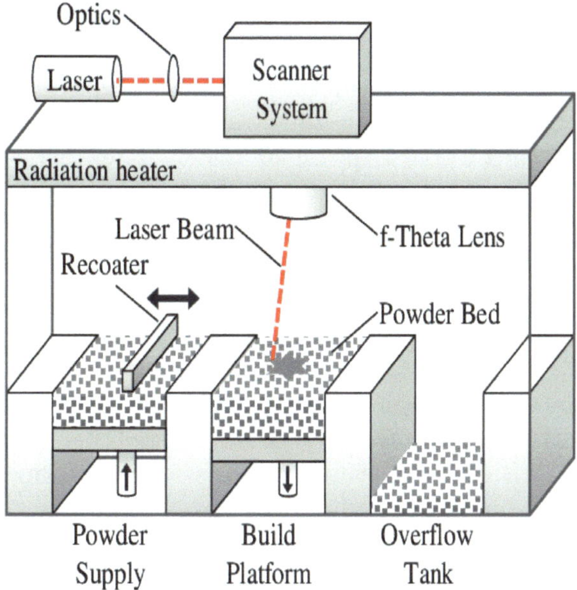

Fig. (10.4). Selective laser sintering.

10.2. DISCUSSION, STRUCTURE AND APPLICATIONS

In 3D printing materials, three types of feed stocks are used such as slurry, powder and bulk based with different physical properties. According to nature of material is to be printed the feed stock is selected. The direct ink writing technique is based on slurry-based feed stock so in this case material which is used to print material is a fluid. The fluids will possess rheological nature such as viscoelastic property with high shear thinning. Generally, in this direct ink writing Ni-Zn ferrite is used as a ceramic suspension because it avoids deformation by gravitational slumping forces. The Ni-Zn ferrite possess spinal ferrite structure, the density of material of Ni-Zn at 930°C is 5.18 g/c.c. The shear thinning exponent varies or ranges from 0 to 1. It shows that all Ni-Zn based slurry exhibits shear thinning characteristics and by measuring complex modulus viscoelastic properties can be analyzed. Ni-Zn based materials can be used as very good magnetic material which can be magnetized and possess permeability [4, 18].

In digital light processing [DLP] the slurry is used to print materials in this technology we can print materials such as mainly whiskers which have honeycomb structure to filter water or oil from multi conditions. To prepare whisker or mesh with holes dense alumina slurry is used to achieve mechanical strength, and borate is coated on surface to achieve further mechanical strength. The coating on the whiskers is super hydrophilic which absorbs the water and percolates into deep due to gravity. On the other side if oil is come into contact with water spread area pressure is increased due to superoleophobicity of oil on whiskers, they prevent the flow of oil with water in mesh. In oil rich, water and oil mixture can be separated by absorption of water. The absorption of water is by sponge ceramics which is multifold crossed struts with highly channeled and ordered structures. This honeycomb structures are very good catalysts for environmental control and as well as energy generation and conservation concerned applications. The polylactic acid is biodegradable plastic which is plant-based origin and is printed into 3D templates using fused deposition machine system [4, 19 - 25]. From digital light processing we can prepare SiBCN ceramic compounds with complex shape, by preceramic polymer. To prepare complex ceramic parts the polymer derived ceramics [PDCs] parts are alternative. The advantages of PDCs are preceramic polymer can be formed into desired form by polymer processing technique, which increases the range of shaping ceramics. At lower temperature the preceramic material can be converted into ceramic polymer than conventional method of powder sintering method. SiBCN ceramics exhibits high thermal stability and oxidation resistance too. It retains amorphous state and hinders crystallization up to 1700°C which implies unique phase stability [4, 19, 23].

In ceramics some metallic frames can be introduced to improve penetration resistance, to form the hybrid structure and to blast loading or compressive loading. 6061-T6 aluminum panels with sandwich corrugated cores are not compatible with monolithic alloy panels with same dimensions. But the recent study suggested that without inserting and triangular corrugated cores exhibited same kind of penetration resistance than high speed spherical projectile of 8.4 g [20]. The 3D printing of porous ceramics has more advantages over conventional porous methods, in conventional method time consumption is more; it needs highly toxic organic materials [solvents], not possible to remove complete residual particles, random porous texture and thin structure. The porous ceramics have many applications in the field of surgical tools, patient specific prostheses and porous ceramic filters fabrication. The pore size distribution and porosity that is microstructure meets different potential advantages. The porous ceramics can be fabricated by different methods such as burning of polymeric sponge impregnated with ceramic slurry, a sol-gel process, sintering and gel casting process [21]. The ceramic membranes which are also porous in nature, made up of clay powder. The most useful clay is kanakara clay. The chemical composition of kanakara clay consists of SiO_2, Al_2O_3, K_2O, Fe_2O_3, CaO, TiO_2, MgO, MnO, P_2O_5, and LO. The ceramic powder particles vary in size; they are 75, 150 and 250 μm size. These membranes are used to filter contaminated water comprising organic and inorganic effluents, so drinking of such water may leads to some water born diseases. The pure water in rural side is difficult to get as they are contaminated by such effluents and organic and inorganic waste. So, the ceramic membranes are very useful in the above said connection, if the size of the particle is minimum then the membrane is most advantages to filter the water [22]. The porous bio glasses are used in the field medicine to replace bones, these glasses are simple ternary compounds made up of SiO_2-CaO-P_2O_5 and they possess sufficient osteogenesis capability. Glass will have unique properties such as apatite layer formation and bone bonding; we call it as bioactivity of bioactive material [24].

CONCLUSIONS

For rising population, human needs also increase in rapid manner. To meet selfish man needs the science and technology is only supplementary which provides needs and luxuries as well. The conventional manufacturing goods do not meet the demands of man so the new technology emerged out of our scientist's efforts, called 3D printing. The conventional manufacturing requires more labor to manufacture. But this 3D printing can manufacture more units within prescribed time, labor and possess more advantage which is low cast which helps to meet all sections of society.

CONSENT FOR PUBLICATION

Not applicable.

CONFLICT OF INTEREST

The authors confirm that this chapter content has no conflict of interest.

ACKNOWLEDGEMENTS

The authors express thankfulness to Dr. P. Sreeramulu, Assistant Professor (English), GITAM, Bangalore for providing English language editing services to this manuscript.

REFERENCES

[1]   H.A. Balogun, R. Sulaiman, S.S. Marzouk, A. Giwa, and S.W. Hasan, "3D printing and surface imprinting technologies for water treatment: A review", *J. Water Process Eng.,* vol. 31, p. 100786, 2019.
[http://dx.doi.org/10.1016/j.jwpe.2019.100786]

[2]   Y. Ben, L. Zhang, S. Wei, T. Zhou, Z. Li, H. Yang, Y. Wang, and C. Selim, "Wong, Chen, H. PVB modified spherical granules of β-TCP by spray drying for 3D ceramic printing", *J. Alloys Compd.,* vol. 721, pp. 312-319, 2017.
[http://dx.doi.org/10.1016/j.jallcom.2017.06.022]

[3]   N. Shahrubudin, and T.C. Lee, "andRamlan, R. An Overview on 3D Printing Technology: Technological, Materials, and Applications", *Procedia Manufacturing,* vol. 35, pp. 1286-1296, 2019.
[http://dx.doi.org/10.1016/j.promfg.2019.06.089]

[4]   P. Biswas, S. Mamatha, S. Naskar, Y.S. Rao, R. Johnson, and G. Padmanabham, "3D extrusion printing of magnesium aluminate spinel ceramic parts using thermally induced gelation of methyl cellulose", *J. Alloys Compd.,* vol. 770, pp. 419-423, 2018.
[http://dx.doi.org/10.1016/j.jallcom.2018.08.152]

[5]   I. Buj-Corral, A. Bagheri, and A. Domínguez-Fernández, "Influence of Structure Support Printing Parameters on Surface Finish of PLA Hemispherical Cups for Emulation of Ceramic Hip Prostheses", *Procedia CIRP,* vol. 68, pp. 347-351, 2018.
[http://dx.doi.org/10.1016/j.procir.2017.12.093]

[6]   Z. Chen, Z. Li, J. Li, C. Liu, C. Liu, Y. Li, P. Wang, H. Yi, C. Lao, and F. Yuelong, "3D printing of ceramics: A review", *J. Eur. Ceram. Soc.,* vol. 31, pp. 661-687, 2019.
[http://dx.doi.org/10.1016/j.jeurceramsoc.2018.11.013]

[7]   H. Chen, X. Wang, F. Xue, Y. Huang, K. Zhou, and D. Zhang, "3D printing of SiC ceramic: Direct ink writing with a solution of preceramic polymers", *J. Eur. Ceram. Soc.,* vol. 38, pp. 5294-5300, 2018.
[http://dx.doi.org/10.1016/j.jeurceramsoc.2018.08.009]

[8]   Z. Chen, D. Zhang, E. Peng, and J. Ding, "3D-Printed Ceramic Structures with *In situ* Grown Whiskers for Effective Oil/Water Separation", *Chem. Eng. J.,* vol. 373, pp. 1223-1232, 2019.
[http://dx.doi.org/10.1016/j.cej.2019.05.150]

[9]   Y. Fu, Z. Chen, G. Xu, Y. Wei, and C. Lao, "Preparation and stereolithography 3D printing of ultralight and ultrastrongZrOC porous ceramics", *J. Alloys Compd.,* vol. 789, pp. 867-873, 2019.
[http://dx.doi.org/10.1016/j.jallcom.2019.03.026]

[10]   R. He, W. Liu, Z. Wu, D. An, M. Huang, H. Wu, and Z. Xie, "Fabrication of complex-shaped zirconia

ceramic parts *via* a DLP- stereolithography-based 3D printing method", *Ceram. Int.,* vol. 44, no. 3, pp. 3412-3416, 2018.
[http://dx.doi.org/10.1016/j.ceramint.2017.11.135]

[11]   S. Hu, Liu, Zhang, Y.,Xue., Z., Wang, Z., Zhoua, Z., Lu, C., Li, H.,Shiwei Wang, S., "3D printed ceramic phosphor and the photoluminescence property under blue laser excitation", *J. Eur. Ceram. Soc.,* vol. 39, pp. 2731-2738, 2019.
[http://dx.doi.org/10.1016/j.jeurceramsoc.2019.03.005]

[12]   T. Wloka, S. Czich, M. Kleinsteuber, E. Moek, C. Weber, M. Gottschaldt, K. Liefeith, and U.S. Schubert, "Microfabrication of 3D-Hydrogels *via* two-photon polymerization of poly(2-ethyl-2-oxazoline) diacrylates", *Eur. Polym. J.,* vol. 122, p. 109295, 2020.
[http://dx.doi.org/10.1016/j.eurpolymj.2019.109295]

[13]   F. Fina, A. Goyanes, C.M. Madla, A. Awad, S.J. Trenfield, J.M. Kuek, P. Patel, S. Gaisford, and A.W. Basit, "3D printing of drug-loaded gyroid lattices using selective laser sintering", *Int. J. Pharm.,* vol. 547, no. 1-2, pp. 44-52, 2018.
[http://dx.doi.org/10.1016/j.ijpharm.2018.05.044] [PMID: 29787894]

[14]   S.F. Barakh Ali, E.M. Mohamed, T. Ozkan, M.A. Kuttolamadom, M.A. Khan, A. Asadi, and Z. Rahman, "Understanding the effects of formulation and process variables on the printlets quality manufactured by selective laser sintering 3D printing", *Int. J. Pharm.,* vol. 570, p. 118651, 2019.
[http://dx.doi.org/10.1016/j.ijpharm.2019.118651] [PMID: 31493496]

[15]   C. Wei, H. Gu, Z. Sun, D. Cheng, Y-H. Chueh, X. Zhang, and L. Li, "Ultrasonic material dispensing-based selective laser melting for 3D printing of metallic components and the effect of powder compression", *Additive Manufacturing,* vol. 29, p. 100818, 2019.
[http://dx.doi.org/10.1016/j.addma.2019.100818]

[16]   X. Shu, and R. Wang, "Thermal residual solutions of beams, plates and shells due to laminated object manufacturing with gradient cooling", *Compos. Struct.,* vol. 174, pp. 366-374, 2017.
[http://dx.doi.org/10.1016/j.compstruct.2017.04.060]

[17]   S. J. Raykar, and Addona D., "Selection of best printing parameters of fused deposition modeling using VIKOR", *Materials Today: Proceedings,* 2019.
[http://dx.doi.org/10.1016/j.matpr.2019.11.104]

[18]   T. An, K-T. Hwang, J-H. Kim, and J. Kim, "Extrusion-based 3D direct ink writing of Ni Zn-ferrite structures with viscoelastic ceramic suspension", *Ceram. Int.,* vol. 46, pp. 6469-6476, 2020.
[http://dx.doi.org/10.1016/j.ceramint.2019.11.127]

[19]   Y. Shushan, S. Dieter, L. Xin, K. Jean-Pierre, and P. Peter Van, "Bart Van der B, 3D printed chemically and mechanically robust membrane by selective laser sintering for separation of oil/water and immiscible organic mixtures", *Chem. Eng. J.,* vol. 385, p. 123816, 2020.
[http://dx.doi.org/10.1016/j.cej.2019.123816]

[20]   X. An, C. Tian, Q. Sun, and Y. Dong, "Effects of material of metallic frame on the penetration resistances of ceramic-metal hybrid structures", *Defence Technology,* vol. 16, pp. 77-87, 2020.
[http://dx.doi.org/10.1016/j.dt.2019.04.015]

[21]   L.C. Hwa, S. Rajoo, A.M. Noor, N. Ahmad, and M.B. Uday, "Recent advances in 3D printing of porous ceramics: A review", *Curr. Opin. Solid State Mater. Sci.,* vol. 21, no. 6, pp. 323-347, 2017.
[http://dx.doi.org/10.1016/j.cossms.2017.08.002]

[22]   L.C. Hwa, M.B. Uday, N. Ahmad, A.M. Noor, S. Rajoo, and K.B. Zakaria, "Integration and fabrication of the cheap ceramic membrane through 3D printing technology", *Materials Today Communications,* vol. 15, pp. 134-142, 2018.
[http://dx.doi.org/10.1016/j.mtcomm.2018.02.029]

[23]   S. Li, W. Duan, T. Zhao, W. Han, L. Wang, R. Dou, and G. Wang, "The fabrication of SiBCN ceramic components from preceramic polymers by digital light processing (DLP) 3D printing technology", *J. Eur. Ceram. Soc.,* vol. 38, no. 14, pp. 4597-4603, 2018.

[http://dx.doi.org/10.1016/j.jeurceramsoc.2018.06.046]

[24]　L. Li, H. Hu, Y. Zhu, M. Zhu, and Z. Liu, "3D-printed ternary SiO_2CaO P_2O_5 bioglass-ceramic scaffolds with tunable compositions and properties for bone regeneration", *Ceram. Int.,* vol. 45, pp. 10997-11005, 2019.
[http://dx.doi.org/10.1016/j.ceramint.2019.02.183]

[25]　S. Mamatha, P. Biswas, D. Das, and R. Johnson, "Fabrication of complex shaped ceramic articles from 3D printed polylactic acid templates by replication process", *Ceram. Int.,* vol. 45, pp. 19577-19580, 2019.
[http://dx.doi.org/10.1016/j.ceramint.2019.06.203]

CHAPTER 11

Advanced Ceramics for Biomedical Applications

B. Venkata Shiva Reddy[1], N. Suresh Kumar[2], K. Chandra Babu Naidu[1,*], Anish Khan[3], Abdullah M. Asiri[3], A. Ratnamala[4] and Sannapaneni Janardan[4]

[1] *Department of Physics, GITAM Deemed to be University, Bangalore-562163, Karnataka, India*

[2] *Department of Physics, JNTUA, Anantapuramu-515002, A.P, India*

[3] *Chemistry Department, Faculty of Science & Center of Excellence for Advanced Materials Research, King Abdulaziz University, P.O. Box 80203, Jeddah, 21589, Saudi Arabia*

[4] *Department of Chemistry, GITAM Deemed to be University, Bangalore-562163, Karnataka, India*

Abstract: Bio ceramics are being prepared by advanced 3D printing technology which is a very good economical technique. Different inorganic materials can be used to print bio ceramics for different applications such as alumina, zirconia, Leucite, lithium disilicate and mica-based ceramics. The main implications of bio ceramics are in dentistry and orthopedics.

Keywords: Alumina, Bio Ceramics, Dentistry, 3D printing, Orthopedics, Zirconia.

11.1. INTRODUCTION

Ceramics played a pivotal role in the civilization of man since time immemorial. Before ceramics invention, man was nomadic for his food because of storage specialties were not in existence. In later period, these ceramics were used as a storage appliance to store their food materials, so ceramic appliances stop nomadic life of man and settlement took place in his desired places. Now ceramic applications pervaded into all spheres of human life from pottery to medicine domain. As a result of scientific research in the field of ceramics from the last five decades, bio ceramics came into medical field. The ceramics intending to contact with living tissue is called bio ceramics which are used to replace damaged part in a human body, generally in orthopedic and dentistry. These ceramics generally composed of inorganic compounds with ionic and covalent bond. Their mechan-

* **Corresponding author Dr. K. Chandra Babu Naidu:** GITAM Deemed To Be University-Bangalore Campus, Bangalore-562163, Karnataka, India; E-mail: chandrababu954@gmail.com

ical properties are good in agreement with natural things, such as great compression strength, very low tensile strength, very stiff, brittle and high young's modulus. The surface properties of bio ceramics have more advantages of high wetting properties and high surface tension which helps to accumulation of proteins, cells and other biological moieties. Alumina, zirconia, calcium phosphate and glasses and glass ceramics are the good examples of bio ceramics. Now research is going on to develop organic-inorganic composition bio-ceramics called hybrid ceramics. The bio-ceramics are relatively strength, low conductivity of heat and electricity and high melting point, the carbon element is treated as ceramic as it possesses many ceramic properties [1].

The manufacturing of bio ceramics is by the method of 3D-printing or additive manufacturing [AM], in this manufacturing no wastage of ceramic material takes place, but in traditional method called subtractive manufacturing [SM] the wastage of materials happens and also time consuming with more manufacturing cost [2]. Bio ceramics are widely used in dentistry and the most of the properties of these materials similar to natural dental material. Bio-ceramics are broadly classified into four categories such as follows:

Glass based systems mainly comprising silica;

Glass based systems mainly comprising silica with fillers, generally crystalline based;

Crystalline based systems with glass fillers;

Poly crystalline solids.

The glass-based systems contain mainly silicon dioxide with different amounts of alumina, feldspars comprise of aluminosilicates occur in nature with different proportion of potassium and sodium. Glass based systems with fillers consists of large range of glass crystalline ratios and crystal type. Crystalline based systems with glass fillers contains glass infiltrated and partially sintered alumina, first introduced in 1988. Polycrystalline solids prepared by sintering crystals form a dense, air free, glass free poly crystalline structure.

The following materials are commonly used, in order to prepare bio-ceramics such as

Zirconium based ceramics (ZrO_2);

Alumina based ceramics (Al_2O_3);

Leucite based ceramics ($K[AlSi_2O_6]$);

Lithium di silicate glass ceramics ($Li_2Si_2O_5$);

Mica based ceramics.

11.2. ZIRCONIUM BASED CERAMICS

These engineering materials are very much popular because of their mechanical properties: fracture toughness, elastic modulus and wear resistance. Yttrium stabilized zirconia (YSZ) possess good implications by superior combination of mechanical properties YSZ will be used as an implant in human body.

11.3. ALUMINA BASED CERAMICS

Alumina based ceramics shows good strength and hardness with excellent wears resistance properties but low fracture toughness. The strength and toughness of the material can be improved by adding tetragonal Zirconia particles.

11.4. LEUCITE BASED CERAMICS

It is potassium alumina silicate-based material with tetragonal structure at room temperature. At higher temperature about 625°C transition takes place from tetragonal to cubic structure. Leucite is implicated in dentistry.

11.5. LITHIUM DISILICATE BASED CERAMICS

This material also shows a good strength, hardness and excellent wear and tear resistance properties. By virtue of mechanical strength and optical properties they are used in single and multiunit dental restoration such as dental crowns, bridges and veneers.

11.6. MICA BASED CERAMICS

Mica minerals are group of silicate minerals that may be in the layered form. The mechanical properties are depending on specific crystal structure formed by cleavage of planes which are located along the planes [2, 3]. The bio ceramics can be used in tissue engineering and biomedical science and bio active glass are third generation bio material and excellent bio compatibility, degradability and bio activity. The bio active glass can form the bone bond through apatite on the surface layer and induce angiogenesis. And the size of the particle of bio glass are very small and high surface area which gives rise high surface to volume ratio leads to good cell attachment used in drug delivery and also used in soft and hard tissue generation. In order to improve mechanical strength of the bio active glass, other elements can be injected such as graphene and graphene oxide nano particles due to their optical, electrical and mechanical properties [4]. In bio ceramics the nature of surface and surface area plays a vital role in their functions, here to develop good advantages we can employ zwitter ionization. This zwitter ionization develops the surface so that nonbacterial adhesion and resistance to nonspecific protein adsorption. The surface formed by the zwitter ionization is neutral because of number of negative charges are equal to positive charges so that surface become neutral. The non-fouling capability *i.e.* hydrogen layer is formed tightly so that nonspecific protein and bacteria cannot accumulate on the surface [5]. Inert bio ceramics are also good bio compatible in connection with anti-bacterial function, one of the inert bio ceramics is the Si_3N_4. However extensive use of Si_3N_4 will restrict anti-bacterial properties. Hence anti-bacterial inorganic agents will be added. Such agents Generally, silver, zinc and copper which are low toxic and superior chemical activity when compared to organic agents. The ZnO is very significant member as an inorganic anti-bacterial agent because of low cost, high bio activity and nature friendly. ZnO is most bio active against some bacteria such as Staphylococcus aureus, Escherichia coli, Salmonella enteritis and Bacillus cereus. Here most important thing is ZnO releases active oxygen and Zn^{2+} which destroys bacterial structure and kill the bacteria. Still we can develop the ZnO anti-bacterial activity by reducing the size of the ZnO into nano scale [6].

11.7. DISCUSSION AND APPLICATIONS

11.7.1. Alumina Applications in Orthopedics

There are seven forms of alumina, among such alpha alumina or corundum is being used to prepare bioceramics as a good biocompatibility material and good stability. Alpha alumina is the highest oxidative form of aluminum such that no more oxidative reactions occur, which gives more physical and chemical stability. The crystal structure is formed by h. c. p layers of oxygen atoms along with two thirds of octahedral sites occupied by alumina atoms. On the surface layer of OH^{+2} groups chemisorption takes place due to O^{2-} layer then bonding of water molecule or proteins will be formed. Thus, the alumina surface shows high wettability. In addition, certain alloys allow forming fluid in the interface of joints of orthroprostheses. To achieve medical grade applications of alumina (ISO6474), ceramics should have no porosity, fine grain size, high grain size homogeneity and high density. The poor micro structure and high level of impurities leads to low stability of bio ceramic. Alumina exhibits very low degree of cell stimulation in macrophage and lymphocytes; if alumina particles are incorporated in macrophage then excellent chemo tactic response exhibits with low toxic ceramic particles. If ceramics are implanted in soft tissue then alumina particles surrounded by layer of connective tissue. After the implantation of alumina ceramics in bone then fast bone formation will be happened without fibrous interposition along without chemical reaction as the alumina is inert material [7].

The alumina weight percentage and pore volume fraction also play pivotal role in bio compatibility in human body in order to maintain suitable density, grain size and flexural limits. While sintering, pores can be obtained by burning foreign combustible substances. The pore size variation will not exhibit any alterations in mechanical properties but pore volume fraction can exhibit in mechanical properties variations in bio ceramics [8].

11.7.2. Alumina Applications in Dentistry

Alumina is monopolistic in nature in dental technology and holds 55% global market in hip replacement bearings. Alumina played a vital role for many decades in experimental and commercial fields such as crowns, dental implants, abutments and bridges. Presently the alumina in dental technology used as orthodontic brackets, more progressively in advance. The orthodontic brackets are the brackets attached to teeth to which the bracing wires of dental wires are attached. Alumina is very translucent and bright material that's why preferred as aesthetic brackets; translucent alumina existing since 1950's developed by Robert coble and translucence is obtained by minimization of light scattering below threshold which requires porosity below 0.1% and ultra fine pore size and grain size (below 2 μm). We will use polycrystalline alumina rather than single crystal because poly crystal alumina is chiefly available and made by powder injection modeling. The following (Fig. **11.1**) shows the orthodontic bracket [8, 9]

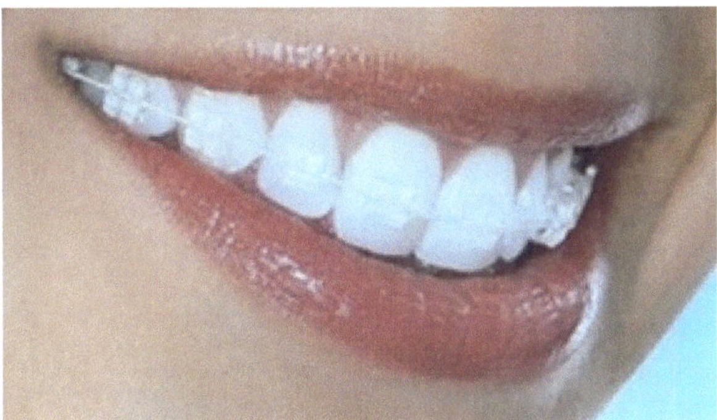

Fig. (11.1). Orthodontic bracket.

In dental crown alumina is used. Dental crown is a veneer tooth made up of pearly white ceramic and color is closely toned to original color of patient's tooth. Mainly in two cases crowns can be used, such as in as cap on broken tooth and cap on abutment of a dental implant. In the case of cap on broken tooth (Fig. **11.2**) the load is concentrated on the broken tooth. In dental crowns sometimes flaws can be formed but reasons not yet identified perfectly, the flaws can be identified by optical microscopy. Alumina is much harder than tooth (alumina strength is 16 GPa. and tooth strength is 4 GPa) that's why alumina can grind down the softer tooth gradually [9, 10].

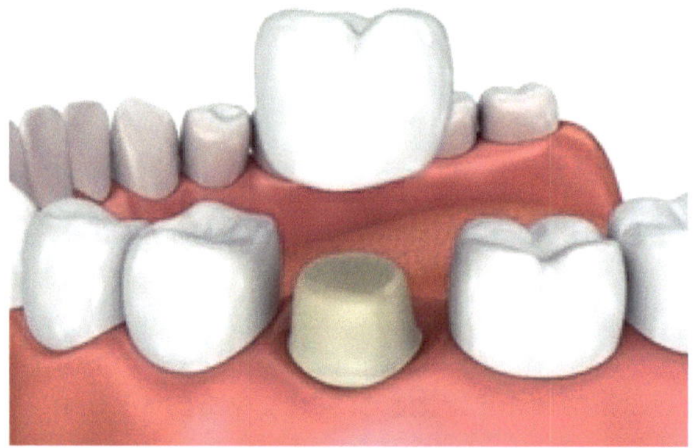

Fig. (11.2). General structure of dental cap.

Dental bridge (Fig. **11.3**) is replacement of damaged teeth or missing teeth that forms a gap in dentition; it involves 3 to 5 elements and is fixed at the extremist by tooth. The dental bridge is older, cheapest and long-established technology. The dental implant is replacement of single tooth using implant supported crown, it is more surgical complex and superior than dental bridge. In the globe nearly 4.5 milion implants have been implanted [10].

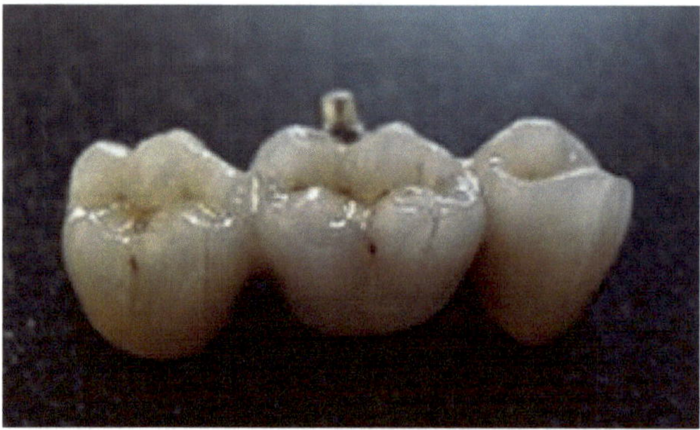

Fig. (11.3). Dental bridge.

11.7.3. Applications of Zirconium

According to Grand view survey the largest producer of zirconium dioxide-based ceramics is the America. Zirconium based materials can be used in bone tissue engineering and orthopedic implants. The extensive research is being taken place in zirconium oxide materials because of its good compression resistance (2000 MPa), good cell culture, cell proliferation, opacity, good fracture strength and biocompatibility *etc.* Zirconia withstand for 50 billion cycles without breaking ceramics. The strength of the bio material surface can be improved by the treatment of zirconium dioxide which can withstand for higher force [11].

11.7.4. Orthopedic Applications

Across the world many natural calamities and accidents are major challenges to the medical field because in these calamities and accidents mainly bone structure defects and bone replacements take place. As a result, a progressive research is being taken place in medicine field. The applications are as follows.

a. Hip Arthroplasty

Hip replacement is expected to increase to around 5 lakh by 2030. The electrochemical behavior of anodic zirconium dioxide in enthused body fluid, zirconia nano tubes provides good stability to the zirconium containing materials which leads to corrosion resistance for hip arthroplasty. Ultrasonically cleaned zirconium foil is subjected to dip etching treatment with $HF/H_2O/HNO_3$ in volume ratio 1:2:4 ratios respectively for a second. One-hour duration anodization is carried out with 20 volts with electrochemical cell in which platinum foil acts as counter electrode with glycerin along with 5 vol % of water and 0.35 MNH4F is taken as a electrolyte. The substrate is dried and subjected to annealing at higher temperature 400°C for three hours duration of time. Then the modified substrate is treated simulated body fluid at 37°C with pH level 7.4 then open circuit voltage is applied and corrosion resistance of bio material is improved. The statistical analysis of the experimental result found better performance of annealed zirconia particles exhibiting stability and corrosive resistance of the substrate. Open circuit potential is 611 ± 25 mV, corrosion potential is -373 ± 31 mV along with corrosion current 210.24 ± 34.76 $nAcm^{-2}$. Anti-corrosive nature and stability of annealed zirconium is due to fine crystalline nature [11, 12].

b. Knee Implants

For degenerative joint disease, total knee arthroplasty is very good treatment in young and middle age patients. However, complications may occur by biological responses to wear particles and local and systematic hypersensitivity reactions triggered by metal ions and such as chromium, molybdenum and cobalt. Zirconium nitrate coated ceramic shows no scratches and delaminating during the wear simulation. Zirconium nitrate coated knee implants avoids release of metal ion in knee joints and reduce wear too [12].

11.8. LEUCITE BASED CERAMICS

Ceramic materials used in dentistry should possess specific properties such as low high mechanical strength, high wear resistance, withstanding for oral environment and similarity with natural tooth. Problem with patient and dental care system is related to failure of fixed prosthetic due to cement dissolution and secondary caries. This leads to bacterial infection and finally removal of prosthesis. The bio active glass on the surface has a specific biological response which forms the bond between tissue and material. The hydraxyapatite (HPa) layer increases the proliferation and increases cell attachment, hence material gap is sealed. The bio active glass materials can be synthesized by sol-gel method [13]. The bottom up approach is most advanced technology to produce nano size (80 nm) leucite crystallites for dispersion in glassy matrix. This nano scale leucite composites only able to produce flexural strength of mean (SD): 109 (10.7) MPa. More recently hydrothermal based preparation super fine leucite ceramics are producing but no strength data is available. The combinations of top down method and crystallization heat treatment also should be considered to yield the good advantages in leucite ceramics [14].

11.9. LITHIUM DISILICATE

Lithium disilicate bio ceramics are being used in dental field as an implantation. Lithium disilicate has more advantages than zirconia in some properties. In crowns of monolithic anatomic configuration where fatigue resistant is more and zirconia ceramics were very susceptible to mouth motion and cyclic load. The lithium disilicate material is used as a glass ceramic because of their bio compatibility and aesthetics. Lithium disilicate glass ceramics core material, such as IPS e.max (IvoclarVivadent, Schaan, Liechtenstein) with high flexural strength about 400 MPa and fracture toughness about 3.0MPa m$^{1/2}$ [15].

11.10. MICA BASED CERAMICS

Varieties of bio ceramic materials were producing extensively in medicine field. Mica is the very good material to produce bio ceramics with high mechanical properties. But for better applications zirconia toughened mica glass are presently used for dental restoration. Glass ceramics and zirconia were common use in dental bio ceramics having limitations of glass ceramics ranging from moderate mechanical properties to the veneer fractures, relative opacity and low temperature degradation of zirconia, hence new material combination must be explored to supersede their drawbacks. Hence, zirconia toughened mica glass ceramic which is indirect restorative bio active ceramic. Dental ceramics can be manufactured in two ways such as monolithic and core veneer bilayered restoration. In monolithic, no thermal capability problems are observed, whereas in the case of veneer bilayered restorations thermal capabilities problem will rise in between core and veneer. Sintering is one of the optimum methods for production of high strength core material. During sintering ceramic is cooled from glass transition temperature to room temperature and in between core veneer interface residual stress is occurred. In such a way stress is developed inside the ceramics so that catastrophic cracks of fractures can be happened in bio ceramics. If thermal expansion co efficient is same for core and veneer then we can avoid the fractures in bioceramics [16].

CONCLUSIONS

In medical field there were many challenges to meet the crisis in connection with dental and bone problem. The main problem was in this field artificial bone and teeth could not match with tissue in order to make interface. The interaction between foreign bodies and human body leads to some infections in host body due to some chemical reaction. For last five decades a good revolution took place in bio ceramics to meet the problems in dental and orthopedic field. Now a days, broken bones and damaged tooth can be easily replaced by this bio ceramics which performs more or less same function as that of natural bones and teeth. But slight differences are there between natural bones and tooth in connection with interfacing between tissue and bone or tooth. Hence further research is essential in the field of bio ceramics to meet the medical challenges.

CONSENT FOR PUBLICATION

Not applicable.

CONFLICT OF INTEREST

The authors confirm that this chapter content has no conflict of interest.

ACKNOWLEDGEMENTS

The authors express thankfulness to Dr. P. Sreeramulu, Assistant Professor (English), GITAM, Bangalore for providing English language editing services to this manuscript.

REFERENCES

[1]   M. Vallet-Regí, and A.J. Salinas, *Ceramics as bone repair materials*. Bone Repair Biomaterials, 2019, pp. 141-178.

[2]   R. Galante, C.G. Figueiredo-Pina, and A.P. Serro, "Additive manufacturing of ceramics for dental applications: A review", *Dent. Mater.*, vol. 35, no. 6, pp. 825-846, 2019.
[http://dx.doi.org/10.1016/j.dental.2019.02.026] [PMID: 30948230]

[3]   M.H. Ghaemi, S. Reichert, A. Krupa, M. Sawczak, A. Zykova, and K. Lobach, "Zirconia ceramics with additions of Alumina for advanced tribological and biomedical applications", *Ceram. Int.*, vol. 43, no. 13, pp. 9746-9752, 2017.
[http://dx.doi.org/10.1016/j.ceramint.2017.04.150]

[4]   K. Ilyas, S. Zahid, M. Batool, A.A. Chaudhry, A. Jamal, and F. Iqbal, "*In-vitro* investigation of graphene oxide reinforced bioactive glass ceramics composites", *J. Non-Cryst. Solids,* vol. 505, pp. 122-130, 2019.
[http://dx.doi.org/10.1016/j.jnoncrysol.2018.10.047]

[5]   I. Izquierdo-Barba, M. Colilla, and M. Vallet-Regí, "Zwitterionic ceramics for biomedical applications", *Acta Biomater.,* vol. 40, pp. 201-211, 2016.
[http://dx.doi.org/10.1016/j.actbio.2016.02.027] [PMID: 26911884]

[6]   T. Liu, X. Zhou, F. Qi, L. Li, Q. Li, and G. Shi, "The effect of T-ZnOw addition on the microstructure, mechanical and antibacterial properties of Si3N4 ceramics for biomedical applications", *Ceram. Int.,* vol. 45, no. 2, pp. 2393-2399, 2018.
[http://dx.doi.org/10.1016/j.ceramint.2018.10.158]

[7]   C. Piconi, and G. Maccauro, *Oxide Ceramics for Biomedical Applications*. Reference Module in Materials Science and Materials Engineering, 2016.
[http://dx.doi.org/10.1016/B978-0-12-803581-8.02151-2]

[8]   K.G.S. Gopinath, S. Pal, and P. Tambe, "Prediction of Weight Percentage Alumina and Pore Volume Fraction in Bio-Ceramics Using Gaussian Process Regression and Minimax Probability Machine Regression", *Materials Today: Proceedings,* vol. 5, no. 5, pp. 12233-12239, 2018.

[9]   M. Øilo, and G.D. Quinn, "Fracture origins in twenty-two dental alumina crowns", *J. Mech. Behav. Biomed. Mater.,* vol. 53, pp. 93-103, 2016.
[http://dx.doi.org/10.1016/j.jmbbm.2015.08.006] [PMID: 26318570]

[10]   A. Ruys, *Dental, tissue scaffold, and other specialized biomedical applications of alumina*. Alumina Ceramics, 2019, pp. 123-137.
[http://dx.doi.org/10.1016/B978-0-08-102442-3.00005-1]

[11]   R. Belli, M. Wendler, J.I. Zorzin, L.H. da Silva, A. Petschelt, and U. Lohbauer, "Fracture toughness mode mixity at the connectors of monolithic 3Y-TZP and LS 2 dental bridge constructs", *J. Eur. Ceram. Soc.,* vol. 35, no. 13, pp. 3701-3711, 2015.
[http://dx.doi.org/10.1016/j.jeurceramsoc.2015.04.022]

[12]   A.L. Puente Reyna, B. Fritz, J. Schwiesau, C. Schilling, B. Summer, P. Thomas, and T.M. Grupp, "Metal ion release barrier function and biotribological evaluation of a zirconium nitride multilayer coated knee implant under highly demanding activities wear simulation", *J. Biomech.,* vol. 79, pp. 88-96, 2018.
[http://dx.doi.org/10.1016/j.jbiomech.2018.07.043] [PMID: 30111498]

[13]   P.H. Kumar, V.K. Singh, A. Srivastava, S.K. Hira, P. Kumar, and P.P. Manna, "Mechanochemically synthesized leucite based bioactive glass ceramic composite for dental veneering", *Ceram. Int.,* vol. 41, no. 9, pp. 11161-11168, 2015.
[http://dx.doi.org/10.1016/j.ceramint.2015.05.065]

[14]   A. Theocharopoulos, X. Chen, R.M. Wilson, R. Hill, and M.J. Cattell, "Crystallization of high-strength nano-scale leucite glass-ceramics", *Dent. Mater.,* vol. 29, no. 11, pp. 1149-1157, 2013.
[http://dx.doi.org/10.1016/j.dental.2013.08.209] [PMID: 24055418]

[15]   Y. Jian, Z.H. He, L. Dao, M.V. Swain, X.P. Zhang, and K. Zhao, "Three-dimensional characterization and distribution of fabrication defects in bilayered lithium disilicate glass-ceramic molar crowns", *Dent. Mater.,* vol. 33, no. 4, pp. e178-e185, 2017.
[http://dx.doi.org/10.1016/j.dental.2017.01.009] [PMID: 28279435]

[16]   S. Gali, and R. K, "Zirconia toughened mica glass ceramics for dental restorations: Wear, thermal, optical and cytocompatibility properties", *Dent. Mater.,* vol. 35, no. 12, pp. 1706-1717, 2019.
[http://dx.doi.org/10.1016/j.dental.2019.08.112] [PMID: 31575490]

CHAPTER 12

Advanced Ceramics for Antimicrobial Applications

M. Prakash[1], N. Suresh Kumar[2] and K. Chandra Babu Naidu[3,*]

[1] *Department of Physics, Sri Krishnadevaraya University, Anantapur-515003, A.P, India*

[2] *Department of Physics, JNTUA, Anantapuramu-515002, A.P, India*

[3] *Department of Physics, GITAM Deemed to be University, Bangalore-562163, Karnataka, India*

Abstract: In this chapter, we specified the origin of ceramics, historical background along with advantagesof the ceramic materials. Meanwhile, the classification of ceramics was also mentioned in a detailed manner. With special interest, we focussed on the antimicrobial applications of advanced ceramics followed by detailed and tabulated data.

Keywords: Antimicrobial Applications, Cell Membrane, Ceramics, Microorganism, Oxidative Stress.

12.1. INTRODUCTION

Generally, the ceramic word came from the Greek language as keramos, and its meaning was "pottery", which was expressed as "to burn" in an ancient language of Sanskrit. Although, the Greeks used the ceramic word to "burnt stuff or "burned earth". This was associated with the function of fire over earthly things. In order to prepare the inorganic and nonmetallic substances, the solid articles were considered based on the art and science of making and usage of the ceramics. According to this, not only the materials like structural clay products, pottery, refractories, porcelain, porcelain enamels, abrasives, cements, and glass, but also nonmetallic magnetic materials, ferroelectrics, and a variety of other products like oxides, carbides and nitrides *etc.*, can be used. United States acquired about ten billion dollars per annum by the ceramic industry. The ceramic industry played a pivotal role in development of many other fields. For instance, refractories are a primary component of the metallurgical industry. Abrasives are crucial to the machine tool and automobile fields. Further, the products of glass ceramics were stipulated to the automobile industry along with electronic and ele-

---

* **Corresponding author Dr. K. Chandra Babu Naidu:** GITAM Deemed To Be University-Bangalore Campus, Bangalore-562163, Karnataka, India; E-mail: chandrababu954@gmail.com

ctrical industries. Cements are needed to the building and architectural industry. Numerous electrical and magnetic ceramics are important to the advancement of computers and electronic control devices. In view of every industrial production and everyday applications of human beings, ceramic materials play significant role. The recently fabricated devices are excluded with ceramic materials owing to their suitable structural, mechanical, electrical, thermal and chemical properties. Finally, the complexity and utility of the materials may exhibit better improvement of their properties in a magnified and tremendous way. In addition to this, many other applications are noticed due to their bigger and greater quality of functioning of the entire system. For better understanding of the ceramics properties the intensive research has been the best path way.

12.2. HISTORICAL PERSPECTIVE

Ceramics were investigated by several archaeologists at the time of 24,000 B.C. Czechoslovakia was first person to find the materials in the form of human and animal figurines, slabs, and balls. The ceramics are made up of the animal fat and one bone combined with other bone ash and the fine clay-like materials. Whenever the ceramics were heated between the range of 500-800°C in domed and horseshoe shaped kilns of partially dug into the ground with walls. During the time of 9,000 B.C, the functional pottery vessels were used in the first time. These were used to store and preserve other food products. Meanwhile, the manufacturing of glass also developed in Upper Egypt at the time of 8,000 B.C. Basically, the pottery holds the calcium oxide (CaO). But during the process of overheating of pottery kiln can be taken in a colored glaze on the ceramic pot. It is known that the ceramic glass preparation was not done comfortably until 1,500 BC. That is, latter, the glass was developed independently using ceramics and fashioned into different items. On the basis of clay-water system, many other traditional ceramics were developed although it was the only first broadly analyzing colloidal system with respect to many advertised ceramic texts. To enable and engineered the ceramic suspensions, the joining of organic processing additives *via* polymer and plasticizers with distinct clay-based system. Moreover, these are the reasons for taking part of the modifications in the rheological behavior and impart handling force to as-formed ceramic texts.

12.3. CLASSIFICATION OF CERAMIC MATERIALS

The Ceramic substances can be classified as follows:

Crystalline ceramics: Which includes silicates, oxides, nonoxides compounds.

Glasses: These are non-crystalline solids which may be silicates or nonsilicates.

Glass-Ceramics: These are primarilymolded as glasses followed by crystallization through controlled heat treatment.

Carbon materials: These include graphite, diamond, fullerenes and nanotubes.

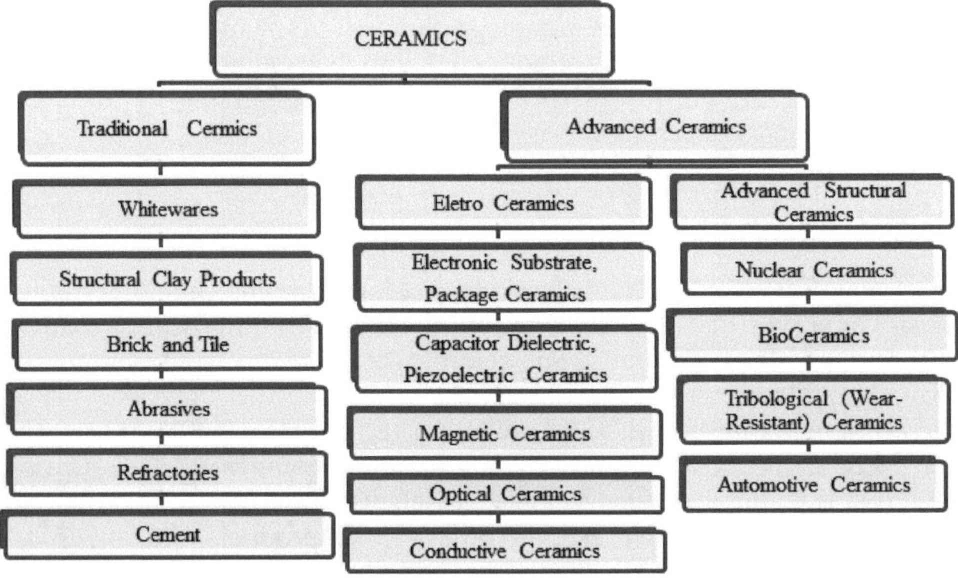

Fig. (12.1). Classification of ceramic materials.

Although Ceramics can also be divided into two parts: (i) Traditional ceramics and (ii) Advanced ceramics, the following (Fig. **12.1**) represents the classification of ceramic materials. The word 'advanced' can be treated as 'technical', 'special', 'fine' or 'engineering'. The traditional ceramics hold a closer association with the previous materials, whereas the first civilization society such as pottery, structural clay products, and clay-based refractories, used the items consisting of cement, concretes and glasses. Eventually, the traditional ceramics played a crucial role in ceramic industry for last few decades and later on, these were concentrated towards the advanced ceramics. The advanced ceramics were developed since last 5 decades. Advanced ceramics consist of ceramics exhibiting the significant electrical, magnetic, electronic, and optical applications. Although traditional ceramics are responsible for the bulk materials, both in tonnage and in dollar volume, a variety of new ceramics have been developed in the last 50 years. These peculiar properties are depending on their special interest with reference to specific requirement in better temperature resistance, greater mechanical properties, distinct electrical properties, and high chemical resistivity. They have been identified more or less suddenly and have become the significant part of the industry. The natural oxide ceramics have been progressed at a great state of uniformity and including the excellent properties as long as accounting for the significant electrical and refractory components. The most oftenly used oxides used are alumina (Al_2O_3), zirconia (ZrO_2), thoria (ThO_2), beryllia (BeO), magnesia (MgO), spinel ($MgAl_2O_4$), and forsterite (Mg_2SiO_4). As long-ago ceramics were in massive part of an empirical art. Ceramic producers were reluctant to replace the details of their processing and manufacturing [1 - 10].

12.4. ANTIMICROBIALS-HISTORICAL OVERVIEW

The usage antimicrobial agents were started practically by ancient Greeks, Indian cultures and Chinese. In general, they prevent the infections with the help of natural sources like plants and molds. Even though plants and molds are organic-sources, to afford antibacterial activity, inorganic-antibacterial agents were also utilized. The silver is an inorganic source which is employed as an antimicrobial-agent for treatment of eye diseases *i.e.*, it prevents the growth of Neisseria gonorrhoeae. This is considered as one of the oldest (documented in the year 1884) documented examples for the medical applications. Alexander Fleming, Ernst Chain and Howard Florey discovered that molds, in particular the mold named Penicillium notatum releases a substance which can prevent the growth of the bacteria and it forms the mold. After this,officiallyit was named as penicillinandthis event is considered as the foundationof "era of antibiotics". The development of naturally obtained antibiotics which are extricated from fungi was trailed thru the advancement of massive synthetic-analogs which were accomplished to have similar or better result. This breakthrough ought to considerable influence on health of the human, because it cured serious ailments like gangrene, syphilis, tuberculosisetc [11 - 16].

The advancement of antibiotics was being subsequent to the growth of their considerable side effects due to bacterial resistance. The bacteria can transform by continuous stress produced through exposure to antimicrobial-agents. These transformations were caused to the development of new strives with a high efficiency for optimization of multidrug resistance.Therefore, there is a necessity for the discovery of new varietyof antibiotics. As far as the numbers of freshly advanced antibiotics are suitable for the utilize as secure, the low-toxic medications are decreased. In the twentieth century, transferable diseases are the main reasons the world wide deaths. In the 21st century of progressive technologies, humans are encountered with transferrable disorders which are still challenging problem. Contamination by critical pathogens throughout regular processes. At present, globally flu-pandemics were made *via* totally new strains which were overall infectious, toxic, completed with huge number of fatal cases. In addition to this, some of the sequences of large indications are as mentioned above. Apart from that to know the in depth of their need, we must be gone through the rigorous research in the area of fighting against pathogens [17 - 24]. Some of the ceramic materials and other types of materials of their antimicrobial applications are given below in (Table **12.1**).

Table. 12.1 Various ceramic materials used in antimicrobial activity.

Material	Antimicrobial Mechanisms	Applications	Microorganism Tested
TiO_2 nanoparticles	(i) Oxidative stress (ROS)	Water purification [69] Self-disinfecting coatings [50,70] Antibacterial textiles [49]	*Escherichia coli* [33, 34, 57, 64, 65] *Staphylococcus aureus* [33,34] MRSA *Staphylococcus aureus* [33] *Pseudomonas aeruginosa* [34] *Enterococcus faecium* [34] *Micrococcus luteus* [33] *Aeromonas hydrophilia* [57] *Poliovirus* 1 [63] *Hepatitis B* [48] *MS-2 phage* [65] *Candida albicans* (fungus) [34]
ZnO nanoparticles	(i) Cell membrane disruption (ii) Release of heavy metal ions (iii) Oxidative stress (ROS)	Deodorant [50] Self-cleaning glass and ceramics [50] *Escherichia coli* [67] *Pseudomonas aeruginosa* [67] Bacillus subtilis [67] *Staphylococcus aureus* [67]	*Escherichia coli* [35] *Pseudomonas aeruginosa* [35] Bacillus subtilis [35] *Staphylococcus aureus* [35] *Escherichia coli* [37, 38] MRSA *Staphylococcus aureus* [66]

(Table 12.1) cont.....

Magnesium oxide nanoparticles	(i) Oxidative stress (ROS) (ii) pH increase	Antimicrobial aqueous solution [62]	*Enterococcus faecalis* [66] *Bacillus subtilis var. niger* [36] *Staphylococcus aureus* [36, 38] *Candida albicans* [66] *Escherichia coli* [39-42] *Staphylococcus aureus* [39] PC12 (cell line) [42]
Graphene and derivatives	(i) Cell membrane disruption (ii) Oxidative stress (ROS independent) (iii) Wrapping	Antibacterial paper [57] Antibacterial fabrics [58] Wound healing [59]	*Escherichia coli* [43 - 45] *Lactobacillus acidophilus*[44] *Bifidobacterium adolescentis* [44]
Carbon nanotubes	(i) Cell membrane disruption (ii) Oxidative stress (ROS and ROS-independent) (iii) Wrapping	Antibacterial coating for implants [60]	*Enterococcus faecalis* [44] *Staphylococcus aureus* [44] *Escherichia coli* [46] *Bacillus subtilis* [47]
Buck minster fullerenes	(i) Oxidative stress (ROS and ROS-independent) (i) Clustering	NA	*MS2 phage* [48] *Escherichia coli* [33, 34] Bacillus subtilis [33, 34]
Nano diamonds		NA	*Escherichia coli* ([38, 42]; [39]; [36, 37]) *Pseudomonas fluorescens* [39] *Salmonella typhimurium* ([39]; [36]) *Vibrio parahaemolyticus* [39]
Chitosan	(i) Cell membrane disruption (ii) Chelation of trace metals (iii) RNA, protein synthesis inhibition	Antibacterial membrane [38] Food preservative [37] Fungicidal and insecticidal (agriculture) [35]	*Staphylococcus aureus* [36, 37, 38, 42] *Listeria monocytogenes* [39] *Bacillus megaterium* [39] *Bacillus cereus* (No. 2002; [37]) *Lactobacillus plantarum* [39] *Lactobacillus brevis* [39] *Botrytis cinerea (fungus)* [37] *Fusarium oxysporum (fungus)* [37] *Drechstera sorokiana (fungus)* [37] *Escherichia coli* [1, 4, 6, 7, 11] *Staphylococcus aureus* [1, 4, 11] *Enterococcus faecalis* [44] *Pseudomonas aeruginosa* [44]

(Table 12.1) cont.....

Nano silver	(i) Release of heavy metal ions (ii) Interrupted trans membrane electron/ion transport (iii) Membrane disruption (iv) Oxidative stress (ROS)	Antibacterial coating for implants [3] Membranes for water purification [40] Antibacterial urinary catheters [5] Antibacterial bone cement [43]	Staphylococcus epidermis [44] *Staphylococcus aureus* (MRSA) [44] *Enterococcus faecium* (VRE) [44] *Klebsiella pneumonia (ESBLpositive)* [44] *Bacillus subtilis* [11]
Copper nanoparticles	(i) Release of heavy metal ions (ii) Cell membrane disruption (iii) Oxidative stress (ROS)	Antifungal coating [25] Water purification [26]	*Escherichia coli* [25-27] *Bacillus subtilis* [26] *Staphylococcus aureus* [25, 27, 61] Staphylococcus epidermis [60] *Candida albicans* [61] *Candida parapsilosis* [61]
Iron nanoparticles	(i) Cell membrane disruption (ii) Oxidative stress (ROS)	Water purification [62]	*Escherichia coli* [28-30] *Bacillus subtilis* [62] *Pseudomonas fluorescens* [62] *Microcystis aeruginosa* [31] *MS-2 phage (virus)* [32]

CONCLUSIONS

Herein, the chapter concludes the in depth study about ceramics and its applications regarding to the various fields including the classification of ceramics. The natural oxide ceramics have been progressed at a great state of uniformity including excellent properties as long as accounting for the significant electrical and refractory components. The oxides used are alumina (Al_2O_3), zirconia (ZrO_2), thoria (ThO_2), beryllia (BeO), magnesia (MgO), spinel

($MgAl_2O_4$), and forsterite (Mg_2SiO_4) suggested for the antimicrobial applications. Mainly, the sustainable advancement of novel antibiotics or persistent enhancement of antimicrobial activity is accessible for antibiotics.As far as the numbers of newly advanced low toxic antibiotics are suitable for secure usage.

CONSENT FOR PUBLICATION

Not applicable.

CONFLICT OF INTEREST

The authors confirm that this chapter content has no conflict of interest.

ACKNOWLEDGEMENTS

The authors express thankfulness to Dr. P. Sreeramulu, Assistant Professor (English), GITAM, Bangalore for providing English language editing services to this manuscript.

REFERENCES

[1]   D.W. Johnson Jr, "Nonconventional powder preparation techniques", *Am. Ceram. Soc. Bull.,* vol. 60, p. 221, 1981.

[2]   D.W. Johnson Jr, "in: "Advances in Powder Technology." ed. G. Y. Chin, American Society for Metals, Metals Park, OH,1982, 22. A. Roosen and H. Hausner, Techniques for agglomeration control during wet-chemical powder synthesis", *Adv. Ceram. Mater.,* vol. 3, p. 131, 1988.

[3]   J.T.G. Overbeek, *Colloid Science.,* Krwyt H. R, Ed., vol. 2. Elsevier: New York, 1952, pp. 4-6.

[4]   H. B. G. Casimier, *Phys. Rev.,* vol. 73, p. 360, 1948.
[http://dx.doi.org/10.1103/PhysRev.73.360]

[5]   T.W. Healy, and D.W. Fuerstenau, "The oxide-water interface-Interrelation of the zero point of charge and the heat of immersion", *J. Colloid Sci.,* vol. 20, pp. 376-386, 1965.
[http://dx.doi.org/10.1016/0095-8522(65)90083-8]

[6]   T.W. Healy, A.P. Herring, and D. Fuerstenau, "The effect of crystal structure on the surface properties of a series of manganese dioxides", *J. Colloid Interface Sci.,* vol. 2J, p. 435, 1966.
[http://dx.doi.org/10.1016/0095-8522(66)90008-0]

[7]   E.J.W. Verwey, and J.T.G. Overbeek, "Theory of stability of lyophobic colloids" (Elsevier, New York,1948). A. E. Lewis, Polar-screen theory for the deflocculation of suspensions", *J. Am. Ceram. Soc.,* vol. 44, p. 233, 1961.

[8]   R.K. McGray, "Mechanical packing of spherical particles", *J. Am. Ceram. Soc.,* vol. 44, p. 513, 1961.
[http://dx.doi.org/10.1111/j.1151-2916.1961.tb13716.x]

[9]   *Nanostructures for Antimicrobial Therapy.* Marija Vukomanovic Jozef Stefan Institute: Ljubljana, Slovenia, 2016.

[10]   "Medical uses of gold compounds: past, present and future, Fricker, S.P", *Gold Bull.,* vol. 29, pp. 53-60, 1996.
[http://dx.doi.org/10.1007/BF03215464]

[11]   "Silver as a disinfectant, Silvestry-Rodriguez, N., Sicairos-Ruelas, E.E., Gerba, C.P., Bright, K.R",

Rev. Environ. Contam. Toxicol., vol. 191, pp. 23-45, 2007.

[12]   A.J. Huh, and Y.J. Kwon, "Nanoantibiotics: A new paradigm for treating infectious diseases using nanomaterials in the antibiotics resistant era", *J. Control. Release,* vol. 156, no. 2, pp. 128-145, 2011.
[http://dx.doi.org/10.1016/j.jconrel.2011.07.002] [PMID: 21763369]

[13]   K.J Simmons, I Chopra, and C.W.G Fishwick, "Structure-based discovery of antibacterial drugs,Simmons", *Nat. Rev,* vol. 8, pp. 501-510, 2000.

[14]   K.J Simmons, I Chopra, and C.W.G Fishwick, "Structure-based discovery of antibacterial drugs", *Natl. Rev.,* vol. 8, pp. 501-510, 2010.

[15]   Q.L. Feng, J. Wu, G.Q. Chen, F.Z. Cui, T.N. Kim, and J.O. Kim, "A mechanistic study of the antibacterial effect of silver ions on *Escherichia coli* and *Staphylococcus aureus*", *J. Biomed. Mater. Res.,* vol. 52, no. 4, pp. 662-668, 2000.
[http://dx.doi.org/10.1002/1097-4636(20001215)52:4<662::AID-JBM10>3.0.CO;2-3]   [PMID: 11033548]

[16]   Q. Li, S. Mahendra, D.Y. Lyon, L. Brunet, M.V. Liga, D. Li, and P.J.J. Alvarez, "Antimicrobial nanomaterials for water disinfection and microbial control: Potential applications and implications", *Water Res.,* vol. 42, no. 18, pp. 4591-4602, 2008.
[http://dx.doi.org/10.1016/j.watres.2008.08.015] [PMID: 18804836]

[17]   M. Bosetti, A. Massè, E. Tobin, and M. Cannas, "Silver coated materials for external fixation devices: *In vitro* biocompatibility and genotoxicity", *Biomaterials,* vol. 23, no. 3, pp. 887-892, 2002.
[http://dx.doi.org/10.1016/S0142-9612(01)00198-3] [PMID: 11771707]

[18]   W.L. Chou, D.G. Yu, and M.C. Yang, "The preparation and characterization of silver loading cellulose acetate hollow fiber membrane for water treatment", *Polym. Adv. Technol.,* vol. 16, pp. 600-607, 2005.
[http://dx.doi.org/10.1002/pat.630]

[19]   S. Saint, J.G. Elmore, S.D. Sullivan, S.S. Emerson, and T.D. Koepsell, "The efficacy of silver alloy-coated urinary catheters in preventing urinary tract infection: A meta-analysis", *Am. J. Med.,* vol. 105, no. 3, pp. 236-241, 1998.
[http://dx.doi.org/10.1016/S0002-9343(98)00240-X] [PMID: 9753027]

[20]   Y. Matsumura, K. Yoshikata, S. Kunisaki, and T. Tsuchido, "Mode of bactericidal action of silver zeolite and its comparison with that of silver nitrate", *Appl. Environ. Microbiol.,* vol. 69, no. 7, pp. 4278-4281, 2003.
[http://dx.doi.org/10.1128/AEM.69.7.4278-4281.2003] [PMID: 12839814]

[21]   I. Sondi, and B. Salopek-Sondi, "Silver nanoparticles as antimicrobial agent: A case study on E. coli as a model for Gram-negative bacteria", *J. Colloid Interface Sci.,* vol. 275, no. 1, pp. 177-182, 2004.
[http://dx.doi.org/10.1016/j.jcis.2004.02.012] [PMID: 15158396]

[22]   L. Cioffi, Torsi, N. Ditaranto, G. Tantillo, L. Ghibelli, L. Sabbatini, T. BleveZacheo, M. D'Alessio, P.G. Zambonin, E. "Traversa Copper nanoparticle/polymer composites with antifungal and bacteriostatic properties", *Chem. Mater.,* vol. 17, pp. 5255-5262, 2005.
[http://dx.doi.org/10.1021/cm0505244]

[23]   C. Cubillo, "Pecharromán, E. Aguilar, J. Santarén, J.S. Moya, Antibacterial activity of copper monodispersed nanoparticles into sepiolite, A. Esteban-", *J. Mater. Sci.,* vol. 41, pp. 5208-5212, 2006.
[http://dx.doi.org/10.1007/s10853-006-0432-x]

[24]   J.P. Ruparelia, A.K. Chatterjee, S.P. Duttagupta, and S. Mukherji, "Strain specificity in antimicrobial activity of silver and copper nanoparticles", *Acta Biomater.,* vol. 4, no. 3, pp. 707-716, 2008.
[http://dx.doi.org/10.1016/j.actbio.2007.11.006] [PMID: 18248860]

[25]   C. Lee, J.Y. Kim, W.I. Lee, K.L. Nelson, J. Yoon, and D.L. Sedlak, "Bactericidal effect of zero-valent iron nanoparticles on *Escherichia coli*", *Environ. Sci. Technol.,* vol. 42, no. 13, pp. 4927-4933, 2008.
[http://dx.doi.org/10.1021/es800408u] [PMID: 18678028]

[26]   "coli E., Li Z., Greden K., Alvarez P.J.J., Gregory K.B., Lowry G.V., Adsorbed polymer and NOM

limits adhesion and toxicity of nano scale zerovalent iron to Environ", *Sci. Technol.,* vol. 44, pp. 3462-3467, 2010.
[http://dx.doi.org/10.1021/es9031198]

[27]   M. Auffan, W. Achouak, J. Rose, M.A. Roncato, C. Chanéac, D.T. Waite, A. Masion, J.C. Woicik, M.R. Wiesner, and J.Y. Bottero, "Relation between the redox state of iron-based nanoparticles and their cytotoxicity toward *Escherichia coli*", *Environ. Sci. Technol.,* vol. 42, no. 17, pp. 6730-6735, 2008.
[http://dx.doi.org/10.1021/es800086f] [PMID: 18800556]

[28]   B. Marsalek, D. Jancula, E. Marsalkova, M. Mashlan, K. Safarova, J. Tucek, and R. Zboril, "Multimodal action and selective toxicity of zerovalent iron nanoparticles against cyanobacteria", *Environ. Sci. Technol.,* vol. 46, no. 4, pp. 2316-2323, 2012.
[http://dx.doi.org/10.1021/es2031483] [PMID: 22242974]

[29]   J.Y. Kim, C. Lee, D.C. Love, D.L. Sedlak, J. Yoon, and K.L. Nelson, "Inactivation of MS2 coliphage by ferrous ion and zero-valent iron nanoparticles", *Environ. Sci. Technol.,* vol. 45, no. 16, pp. 6978-6984, 2011.
[http://dx.doi.org/10.1021/es201345y] [PMID: 21726084]

[30]   W. Kangwansupamonkon, V. Lauruengtana, S. Surassmo, and U. Ruktanonchai, "Antibacterial effect of apatite-coated titanium dioxide for textiles applications", *Nanomedicine (Lond.),* vol. 5, no. 2, pp. 240-249, 2009.
[http://dx.doi.org/10.1016/j.nano.2008.09.004] [PMID: 19223243]

[31]   K.P. Kühn, I.F. Chaberny, K. Massholder, M. Stickler, V.W. Benz, H.G. Sonntag, and L. Erdinger, "Disinfection of surfaces by photocatalytic oxidation with titanium dioxide and UVA light", *Chemosphere,* vol. 53, no. 1, pp. 71-77, 2003.
[http://dx.doi.org/10.1016/S0045-6535(03)00362-X] [PMID: 12892668]

[32]   A. Azam, A.S. Ahmed, M. Oves, M.S. Khan, S.S. Habib, and A. Memic, "Antimicrobial activity of metal oxide nanoparticles against Gram-positive and Gram-negative bacteria: a comparative study", *Int. J. Nanomedicine,* vol. 7, pp. 6003-6009, 2012.
[http://dx.doi.org/10.2147/IJN.S35347] [PMID: 23233805]

[33]   L. Huang, D.Q. Li, Y.J. Lin, M. Wei, D.G. Evans, and X. Duan, "Controllable preparation of Nano-MgO and investigation of its bactericidal properties", *J. Inorg. Biochem.,* vol. 99, no. 5, pp. 986-993, 2005.
[http://dx.doi.org/10.1016/j.jinorgbio.2004.12.022] [PMID: 15833320]

[34]   T. Jin, and Y. He, "Antibacterial activities of magnesium oxide (MgO) nanoparticles against foodborne pathogens", *J. Nanopart. Res.,* vol. 13, pp. 6877-6885, 2011.
[http://dx.doi.org/10.1007/s11051-011-0595-5]

[35]   S. Makhluf, R. Dror, Y. Nitzan, Y. Abramovich, R. Jelinek, and A. Gedanken, "Microwave-assisted synthesis of nanocrystalline MgO and its use as a bacteriocide", *Adv. Funct. Mater.,* vol. 15, pp. 1708-1715, 2005.
[http://dx.doi.org/10.1002/adfm.200500029]

[36]   O. Akhavan, and E. Ghaderi, "Toxicity of graphene and graphene oxide nanowalls against bacteria", *ACS Nano,* vol. 4, no. 10, pp. 5731-5736, 2010.
[http://dx.doi.org/10.1021/nn101390x] [PMID: 20925398]

[37]   S. Liu, T.H. Zeng, M. Hofmann, E. Burcombe, J. Wei, R. Jiang, J. Kong, and Y. Chen, "Antibacterial activity of graphite, graphite oxide, graphene oxide, and reduced graphene oxide: membrane and oxidative stress", *ACS Nano,* vol. 5, no. 9, pp. 6971-6980, 2011.
[http://dx.doi.org/10.1021/nn202451x] [PMID: 21851105]

[38]   M. Raffi, S. Mehrwan, T.M. Bhatti, J.I. Akhter, A. Hameed, and W. Yawar, M.M. ul Hasan, "Investigations into the antibacterial behavior of copper nanoparticles against *Escherichia coli*", *Ann. Microbiol.,* vol. 60, pp. 75-80, 2010.

[http://dx.doi.org/10.1007/s13213-010-0015-6]

[39]   Y. Tu, M. Lv, P. Xiu, T. Huynh, M. Zhang, M. Castelli, Z. Liu, Q. Huang, C. Fan, H. Fang, and R. Zhou, "Destructive extraction of phospholipids from *Escherichia coli* membranes by graphene nanosheets", *Nat. Nanotechnol.,* vol. 8, no. 8, pp. 594-601, 2013.
[http://dx.doi.org/10.1038/nnano.2013.125] [PMID: 23832191]

[40]   O. Akhavan, E. Ghaderi, and A. Esfandiar, "Wrapping bacteria by graphene nanosheets for isolation from environment, reactivation by sonication, and inactivation by near-infrared irradiation", *J. Phys. Chem. B,* vol. 115, no. 19, pp. 6279-6288, 2011.
[http://dx.doi.org/10.1021/jp200686k] [PMID: 21513335]

[41]   S. Kang, M. Herzberg, D.F. Rodrigues, and M. Elimelech, "Antibacterial effects of carbon nanotubes: size does matter!", *Langmuir,* vol. 24, no. 13, pp. 6409-6413, 2008.
[http://dx.doi.org/10.1021/la800951v] [PMID: 18512881]

[42]   H. Chen, B. Wang, D. Gao, M. Guan, L. Zheng, H. Ouyang, Z. Chai, Y. Zhao, and W. Feng, "Broad-spectrum antibacterial activity of carbon nanotubes to human gut bacteria", *Small,* vol. 9, no. 16, pp. 2735-2746, 2013.
[http://dx.doi.org/10.1002/smll.201202792] [PMID: 23463684]

[43]   C.D. Vecitis, K.R. Zodrow, S. Kang, and M. Elimelech, "Electronic-structure-dependent bacterial cytotoxicity of single-walled carbon nanotubes", *ACS Nano,* vol. 4, no. 9, pp. 5471-5479, 2010.
[http://dx.doi.org/10.1021/nn101558x] [PMID: 20812689]

[44]   D.Y. Lyon, J.D. Fortner, C.M. Sayes, V.L. Colvin, and J.B. Hughe, "Bacterial cell association and antimicrobial activity of a C60 water suspension", *Environ. Toxicol. Chem.,* vol. 24, no. 11, pp. 2757-2762, 2005.
[http://dx.doi.org/10.1897/04-649R.1] [PMID: 16398110]

[45]   D.Y. Lyon, L.K. Adams, J.C. Falkner, and P.J. Alvarezt, "Antibacterial activity of fullerene water suspensions: effects of preparation method and particle size", *Environ. Sci. Technol.,* vol. 40, no. 14, pp. 4360-4366, 2006.
[http://dx.doi.org/10.1021/es0603655] [PMID: 16903271]

[46]   A.R. Badireddy, E.M. Hotze, S. Chellam, P. Alvarez, and M.R. Wiesner, "Inactivation of bacteriophages *via* photosensitization of fullerol nanoparticles", *Environ. Sci. Technol.,* vol. 41, no. 18, pp. 6627-6632, 2007.
[http://dx.doi.org/10.1021/es0708215] [PMID: 17948818]

[47]   J. Beranová, G. Seydlová, H. Kozak, O. Benada, R. Fišer, A. Artemenko, I. Konopásek, and A. Kromka, "Sensitivity of bacteria to diamond nanoparticles of various size differs in gram-positive and gram-negative cells", *FEMS Microbiol. Lett.,* vol. 351, no. 2, pp. 179-186, 2014.
[http://dx.doi.org/10.1111/1574-6968.12373] [PMID: 24386940]

[48]   J. Wehling, R. Dringen, R.N. Zare, M. Maas, and K. Rezwan, "Bactericidal activity of partially oxidized nanodiamonds", *ACS Nano,* vol. 8, no. 6, pp. 6475-6483, 2014.
[http://dx.doi.org/10.1021/nn502230m] [PMID: 24861876]

[49]   M.E.I. Badawy, E.I. Rabea, and T.M. Rogge, "Fungicidal and insecticidal activity of O-acyl chitosan derivatives", *Polym. Bull.,* vol. 54, pp. 279-289, 2005.
[http://dx.doi.org/10.1007/s00289-005-0396-z]

[50]   L. Qi, Z. Xu, X. Jiang, C. Hu, and X. Zou, "Preparation and antibacterial activity of chitosan nanoparticles", *Carbohydr. Res.,* vol. 339, no. 16, pp. 2693-2700, 2004.
[http://dx.doi.org/10.1016/j.carres.2004.09.007] [PMID: 15519328]

[51]   E.I. Rabea, M.E.T. Badawy, C.V. Stevens, G. Smagghe, and W. Steurbaut, "Chitosan as antimicrobial agent: applications and mode of action", *Biomacromolecules,* vol. 4, no. 6, pp. 1457-1465, 2003.
[http://dx.doi.org/10.1021/bm034130m] [PMID: 14606868]

[52]   T.M. Don, C.C. Chen, C.K. Lee, W.Y. Cheng, and L.P. Cheng, "Preparation and antibacterial test of

chitosan/PAA/PEGDA bi-layer composite membranes", *J. Biomater. Sci. Polym. Ed.,* vol. 16, no. 12, pp. 1503-1519, 2005.
[http://dx.doi.org/10.1163/156856205774576718] [PMID: 16366335]

[53] H.K. No, N.Y. Park, S.H. Lee, and S.P. Meyers, "Antibacterial activity of chitosans and chitosan oligomers with different molecular weights", *Int. J. Food Microbiol.,* vol. 74, no. 1-2, pp. 65-72, 2002.
[http://dx.doi.org/10.1016/S0168-1605(01)00717-6] [PMID: 11929171]

[54] J. Holappa, M. Hjálmarsdóttir, and M. Másson, "Antimicrobial activity of chitosan N-betainates", *Carbohydr. Polym.,* vol. 65, pp. 114-118, 2006.
[http://dx.doi.org/10.1016/j.carbpol.2005.11.041]

[55] F. Hossain, O.J. Perales-Perez, S. Hwang, and F. Román, "Antimicrobial nanomaterials as water disinfectant: applications, limitations and future perspectives", *Sci. Total Environ.,* vol. 466-467, pp. 1047-1059, 2014.
[http://dx.doi.org/10.1016/j.scitotenv.2013.08.009] [PMID: 23994736]

[56] T. Tong, A. Shereef, J. Wu, C.T. Binh, J.J. Kelly, J.F. Gaillard, and K.A. Gray, "Effects of material morphology on the phototoxicity of nano-TiO2 to bacteria", *Environ. Sci. Technol.,* vol. 47, no. 21, pp. 12486-12495, 2013.
[http://dx.doi.org/10.1021/es403079h] [PMID: 24083465]

[57] V. Alt, T. Bechert, P. Steinrücke, M. Wagener, P. Seidel, E. Dingeldein, E. Domann, and R. Schnettler, "An *in vitro* assessment of the antibacterial properties and cytotoxicity of nanoparticulate silver bone cement", *Biomaterials,* vol. 25, no. 18, pp. 4383-4391, 2004.
[http://dx.doi.org/10.1016/j.biomaterials.2003.10.078] [PMID: 15046929]

[58] A. Panacek, L. Kvítek, R. Prucek, M. Kolar, R. Vecerova, N. Pizúrova, V.K. Sharma, T. Nevecna, and R. Zboril, "Silver colloid nanoparticles: synthesis, characterization, and their antibacterial activity", *J. Phys. Chem. B,* vol. 110, no. 33, pp. 16248-16253, 2006.
[http://dx.doi.org/10.1021/jp063826h] [PMID: 16913750]

[59] T. Kruk, K. Szczepanowicz, J. Stefańska, R.P. Socha, and P. Warszyński, "Synthesis and antimicrobial activity of monodisperse copper nanoparticles", *Colloids Surf. B Biointerfaces,* vol. 128, pp. 17-22, 2015.
[http://dx.doi.org/10.1016/j.colsurfb.2015.02.009] [PMID: 25723345]

[60] M. Diao, and M. Yao, "Use of zero-valent iron nanoparticles in inactivating microbes", *Water Res.,* vol. 43, no. 20, pp. 5243-5251, 2009.
[http://dx.doi.org/10.1016/j.watres.2009.08.051] [PMID: 19783027]

[61] R.J. Watts, S. Kong, and M.P. Orr, "Photocatalytic inactivation of coliform bacteria and viruses in secondary wastewater effluent", *Water Res.,* vol. 29, pp. 95-100, 1995.
[http://dx.doi.org/10.1016/0043-1354(94)E0122-M]

[62] L. Zan, W. Fa, T. Peng, and Z.K. Gong, "Photocatalysis effect of nanometer TiO2 and TiO2-coated ceramic plate on Hepatitis B virus", *J. Photochem. Photobiol. B,* vol. 86, no. 2, pp. 165-169, 2007.
[http://dx.doi.org/10.1016/j.jphotobiol.2006.09.002] [PMID: 17055286]

[63] M. Cho, H. Chung, W. Choi, and J. Yoon, "Different inactivation behaviors of MS-2 phage and *Escherichia coli* in TiO2 photocatalytic disinfection", *Appl. Environ. Microbiol.,* vol. 71, no. 1, pp. 270-275, 2005.
[http://dx.doi.org/10.1128/AEM.71.1.270-275.2005] [PMID: 15640197]

[64] A. Monzavi, S. Eshraghi, R. Hashemian, and F. Momen-Heravi, "*In vitro* and *ex vivo* antimicrobial efficacy of nano-MgO in the elimination of endodontic pathogens", *Clin. Oral Investig.,* vol. 19, no. 2, pp. 349-356, 2015.
[http://dx.doi.org/10.1007/s00784-014-1253-y] [PMID: 24859291]

[65] C.F. Haung, Y.H. Chan, and L.K. Chen, "Preparation, characterization, and properties of anticoagulation and antibacterial films of carbon-based nanowires fabricated on surfaces of Ti implants", *J. Electrochem. Soc.,* vol. 160, pp. H392-H397, 2013.

[http://dx.doi.org/10.1149/2.109306jes]

[66]   W. Hu, C. Peng, W. Luo, M. Lv, X. Li, D. Li, Q. Huang, and C. Fan, "Graphene-based antibacterial paper", *ACS Nano,* vol. 4, no. 7, pp. 4317-4323, 2010.
[http://dx.doi.org/10.1021/nn101097v] [PMID: 20593851]

[67]   J. Zhao, B. Deng, M. Lv, J. Li, Y. Zhang, H. Jiang, C. Peng, J. Li, J. Shi, Q. Huang, and C. Fan, "Graphene oxide-based antibacterial cotton fabrics", *Adv. Healthc. Mater.,* vol. 2, no. 9, pp. 1259-1266, 2013.
[http://dx.doi.org/10.1002/adhm.201200437] [PMID: 23483725]

[68]   B. Lu, T. Li, H. Zhao, X. Li, C. Gao, S. Zhang, and E. Xie, "Graphene-based composite materials beneficial to wound healing", *Nanoscale,* vol. 4, no. 9, pp. 2978-2982, 2012.
[http://dx.doi.org/10.1039/c2nr11958g] [PMID: 22453925]

SUBJECT INDEX

www.ingramcontent.com/pod-product-compliance
Lightning Source LLC
Chambersburg PA
CBHW042040110726
48006CB00002B/241